# STRUCTURE AND BONDING

## Volume 21

Editors: J. D. Dunitz, Zürich
P. Hemmerich, Konstanz · R. H. Holm, Cambridge
J. A. Ibers, Evanston · C. K. Jørgensen, Genève
J. B. Neilands, Berkeley · D. Reinen, Marburg
R. J. P. Williams, Oxford

With 62 Figures

Springer-Verlag
Berlin Heidelberg GmbH 1975

ISBN 978-3-662-15532-5    ISBN 978-3-540-37395-7 (eBook)
DOI 10.1007/978-3-540-37395-7

Library of Congress Catalog Card Number 67-11280

© by Springer–Verlag Berlin Heidelberg 1975
Originally published by Springer-Verlag Berlin Heidelberg New York in 1975
Softcover reprint of the hardcover 1st edition 1975

# Contents

STRUCTURE AND BONDING is issued at irregular intervals, according to the material received. With the acceptance for publication of a manuscript, copyright of all countries is vested exclusively in the publisher. Only papers not previously published elsewhere should be submitted. Likewise, the author guarantees against subsequent publication elsewhere. The text should be as clear and concise as possible, the manuscript written on one side of the paper only. Illustrations should be limited to those actually necessary.

---

Manuscripts will be accepted by the editors:

| | |
|---|---|
| Professor Dr. *Jack D. Dunitz* | Laboratorium für Organische Chemie der Eidgenössischen Hochschule CH-8006 Zürich, Universitätsstraße 6/8 |
| Professor Dr. *Peter Hemmerich* | Universität Konstanz, Fachbereich Biologie D-7750 Konstanz, Postfach 733 |
| Professor *Richard H. Holm* | Department of Chemistry, Massachusetts Institute of Technology Cambridge, Massachusetts 02139/U.S.A. |
| Professor *James A. Ibers* | Department of Chemistry, Northwestern University Evanston, Illinois 60201/U.S.A. |
| Professor Dr. *C. Klixbüll Jørgensen* | 51, Route de Frontenex, CH-1207 Genève |
| Professor *Joe B. Neilands* | University of California, Biochemistry Department Berkeley, California 94720/U.S.A. |
| Professor Dr. *Dirk Reinen* | Fachbereich Chemie der Universität Marburg D-3550 Marburg, Gutenbergstraße 18 |
| Professor *Robert Joseph P. Williams* | Wadham College, Inorganic Chemistry Laboratory Oxford OX1 3QR/Great Britain |

---

SPRINGER-VERLAG

D-6900 Heidelberg 1
P. O. Box 105280
Telephone (06221) 487·1
Telex 04-61723

D-1000 Berlin 33
Heidelberger Platz 3
Telephone (030) 822001
Telex 01-83319

SPRINGER-VERLAG
NEW YORK INC.

175, Fifth Avenue
New York, N. Y. 10010
Telephone 673-2660

# The Study of Covalency by Magnetic Neutron Scattering

## Bruce C. Tofield*

Bell Laboratories, Holmdel, New Jersey 07733, U.S.A.

## Table of Contents

---

* Present address: Materials Physics Division, A.E.R.E. Harwell, Oxfordshire OX11 0RA, England

# 1. Introduction

Neutron scattering is a very powerful tool in the investigation of many areas of chemistry and physics (1, 2). This review will deal with the study of spin distributions and bonding in non-metallic solids by elastic magnetic neutron scattering.

The magnetic scattering of neutrons has provided new information on many aspects of magnetic phenomena. A principal contribution has been the determination by elastic Bragg scattering of the magnetic structures of magnetically ordered metals, alloys, semiconductors and insulators. Compounds of the first row transition metals have been principally studied but much application has also been made to rare earth and actinide materials and to mixed materials such as the rare earth iron garnets and rare earth transition metal oxide perovskites. This type of study is analogous to the determination of the crystallographic space groups of crystals by X-ray or nuclear neutron scattering and depends in principle on the positions in reciprocal space of diffraction peaks rather than on precise intensity measurements (although these may also be needed in some instances to distinguish between alternative structures). A large variety of magnetic space groups is observed including helical structures, as well as collinear ones. One- and two-dimensional as well as three-dimensional systems have been investigated. Aspects of this work have been recently described (3).

In an analogous way to the determination of atom positions and the charge distribution in a crystal by measurement of X-ray scattering intensities, the magnetic moment distribution in a magnetically ordered crystal may be found from the measured intensities of magnetic reflections. This is achieved either by considering the magnetic ions to be non-overlapping and determining the effective magnetic moments or, in the case that fairly complete data are collected, by Fourier transform techniques. The latter approach is essential in many metallic systems where the magnetic atoms are not isolated and where conduction band spin density may also be present. Most experiments performed to date on salts have, however, been principally concerned with the details of the magnetic structure; in such cases it is generally sufficient to ascertain that the magnetic moments are reasonably close to the values expected and intensity data of high accuracy are not required. Consequently, detailed determinations of magnetic moments and spin distributions, essential in the study of bonding and which are discussed below, remain fairly few in number. The measurement of covalency effects has in fact often involved a high precision re-investigation of materials of already known magnetic structure.

Covalent interactions in solids change, inter alia, the charge distribution of the valence electrons of atoms relative to the isolated atoms or ions, or to some conceptual extreme such as the ionic model. The charge and spin densities of isolated atoms or ions, at least, are relatively amenable to calculation. For systems containing atoms with unpaired electrons, therefore, bonding effects should be reflected in the magnetic moment distribution and may be studied by magnetic neutron scattering if a careful investigation of the intensity of magnetic scattering is made. The experimental data may be compared either with ab initio calculations or interpreted via a bonding model. This has recently been discussed in general terms for several experimental techniques that give fairly direct information on bonding (4). For many of the results obtained to date by magnetic neutron scattering it has been convenient to make a simple interpretation using the molecular orbital (MO) model of transition metal complexes. This is useful both for relating trends observed from one ligand or metal to the next, and for comparing neutron data with that obtained by magnetic resonance study of ligand hyperfine interactions (LHFI). A number of experiments nevertheless reveal the deficiencies of the simple model, and, especially where fairly complete, accurate spin density data are available, a more sophisticated analysis is necessary.

Some of the most impressive demonstrations of the power of neutron diffraction in the determination of spin density distributions were the studies of the ferromagnetic transition metals iron (5), nickel (6), and cobalt (7, 8) using polarized neutrons. Such collective electron systems, however, are not within the scope of a localized electron description such as the molecular orbital model and we will not be concerned with them here. But most compounds which have been studied so far by neutrons with regard to bonding, are concentrated (in respect of the metal ions with unpaired spin) magnetically ordered systems (generally antiferromagnetic) and it is pertinent to enquire about the applicability of the molecular orbital model in interpreting the data. Criteria for the application of localized or bond models of the valence electrons in oxides and other systems have been discussed by *Goodenough* (9). There is clearly no problem with materials such as TiO, $ReO_3$ or $LaNiO_3$, which are Pauli paramagnetic with wide bands and show no magnetic ordering effects. There is a problem with narrow-band metallic materials which may demonstrate magnetic ordering effects with apparently localized moments below the ordering temperature, but where the magnitude of the spin associated with the metal ion must be interpreted by band theory rather than by a localized electron model of covalency wherein magnetic ordering is the result of superexchange effects[1] (10). Thus $V_2O_3$ (14) and $NiS_2$ (15) possess abnormally low moments (1.2 $\mu_B$ and 1.17 $\mu_B$, respectively) in the antiferromagnetically ordered phases, although the paramagnetic moments are not unusual, and these large moment reductions possibly should not be interpreted

---

[1]) Superexchange effects in localized electron materials, with generally dominant antiferromagnetic interactions are related to covalency effects; a good discussion has been given by *Owen* and *Thornley* (11). Many of the simple magnetic arrangements observed for the materials discussed here may be rationalized from the numbers of unpaired electrons associated with the ions concerned (12). The increase in magnetic ordering temperatures observed with increase of oxidation state (e.g., for $d^5$ ions, MnO has $T_N = 120K$ but $Fe_2O_3$ has $T_N = 948K$) reflects the increase in covalency. But as covalency or overlap increases further and a narrow band description becomes appropriate, ordering temperatures fall rapidly until a Pauli paramagnetic state is reached, e.g., for $d^2$ ions $LaVO_3$ has $T_N = 137K$, $CaCrO_3$ has $T_N = 90K$ and $CaMoO_3$ is Pauli paramagnetic (13).

in the normal manner used for covalent localized-electron materials[2]). Such problems may also arise, for example, for $SrFeO_3$ which, as was anticipated from measurements on oxygen deficient samples (17), has recently been shown to have a helical magnetic structure in the antiferromagnetic phase (18), and for non-transition metal compounds such as US (19), although in the latter case it could be shown that the magnetic electrons are primarily associated with $5f$ and not $6d$ orbitals. However, it is believed (9), that a localized electron model is appropriate to the (mainly) divalent and trivalent oxides and fluorides of $3d^3$, $3d^5$ and $3d^8$ configuration which have had the majority of attention in the study of bonding by magnetic neutron diffraction. This will be assumed to be the case for all the materials discussed below.

The choice of magnetically concentrated materials, where such an ambiguity may exist, has mainly been dictated by experimental reasons. In magnetically ordered systems the scattering is concentrated into discrete directions in the reciprocal lattice and may be readily observed as magnetic Bragg reflections. More dilute systems (either with larger ligands or with the magnetic ion doped into an insulating host as is the case in ESR work) tend to order, if at all, at inconveniently low temperatures, and in the paramagnetic phase, the magnetic scattering is not easily subtracted with any accuracy from other contributions to the background such as nuclear incoherent and thermal diffuse scattering, or from the nuclear Bragg reflections. Apart from one experiment on ruby (20) in which the diffuse scattering at long wavelengths was measured in an attempt to detect spin density transferred to the oxygens, only one measurement of paramagnetic scattering [on the very high spin ($S = 7/2$) material $Gd_2O_3$ measured by polarization analysis (21)] has been presented that is of sufficient accuracy to give information on bonding. Where salts remain paramagnetic to very low temperatures, however, considerable alignment of spins may be obtained by application of a magnetic field making the material effectively ferromagnetic and thus amenable to study by polarized neutrons. Magnetic intensities may then be measured with high accuracy using polarized beam techniques. Only one such study has so far been made, on $K_2NaCrF_6$ (22), but with the operation of high flux reactors and the use of polarization analysis (see below) with higher reflectivity polarizers than have hitherto been used, it is expected that information on paramagnetic systems will provide the main contribution to the study of covalency by magnetic neutron scattering during the remainder of this decade.

Except in the few cases where a fairly large set of data has been collected on single crystals, and the magnetic moment density in the unit cell obtained by

---

[2]) One must also be aware of the complicating effects of nonstoichiometry. Thus, $NiS_2$ studied by *Hastings* and *Corliss* (15) was claimed to be $NiS_{1.94}$ (16), and the effect of sulphur composition on the magnetic properties was shown to be significant. NiS has recently been studied by powder diffraction methods (16a). The ordered moment was found to change from $1.45\mu_B$ to $1.00\mu_B$ at 4.2K as the stoichiometry changed from $Ni_{1.00}S$ to $Ni_{0.97}S$ with only a 0.6% decrease in the Ni—Ni distance. This is fairly conclusive evidence that NiS is a itinerant-electron antiferromagnet with no local moments, and thus discussion of the magnitude of the observed moments is beyond the scope of this review.

Fourier techniques, the derivation of covalency parameters has been achieved by comparison of observed absolute intensities of magnetic reflections of polycrystalline samples with those calculated from a free ion model. In this situation it is assumed that the magnetization density in the crystal is the sum of the magnetic moment densities associated with the individual ions — a reasonable approximation within the localized electron considerations. The covalent redistribution of the charge spreads some spin on to the ligands where, by virtue of the angular dependence of magnetic scattering it does not contribute significantly to the magnetic reflections observed for many commonly found magnetic structures. In fact, for most antiferromagnets, it is quenched entirely. The spin density on the metal ion is thus reduced with respect to the free ion value — the extent of the reduction being a measure of the covalency. In paramagnetic systems the spin density on the ligands is not quenched and may be determined from the LHFI measured by ESR or NMR. Thus, neutron diffraction and the resonance techniques are complementary in the investigation of bonding effects.

The angular variation of scattered magnetic intensity (given by the magnetic form factor) is related to the radial distribution of the magnetic electrons. In most of the covalency experiments using powders, it was not possible to measure a sufficient number of magnetic reflections to draw any conclusions about the radial distribution of the $3d$ electrons compared to the free ion model (measurements of the magnetic moment are generally made with the lowest angle magnetic reflection [which often has $\sin\theta/\lambda \approx 0.1$ Å$^{-1}$] to provide the highest accuracy and to minimize uncertainties in the radial distribution). But accurate measurement of form factors has been possible in some experiments on single crystals and in one or two cases of interest with powders. Such data provide direct evidence on the radial behavior of the $d$ orbitals in transition metal salts, and throw light, for example, on the nephelauxetic effect (radial expansion of the metal valence orbitals due to overlap with the ligands, and consequent screening from the nuclear charge of the metal) observed in electronic spectroscopy (23) as a reduction in the Coulombic repulsions between $d$ electrons. Some experiments are discussed below.

Study has been almost entirely restricted to systems with spin-only ground states as the interpretation is much simpler than for systems with orbitally degenerate ground states. But even if experimental moment reductions and form factors are measured with good accuracy it should be noted that some pitfalls may attend a simple analysis using the molecular orbital model. For antiferromagnets, zero-point motion effects cause a reduction in the observed spin regardless of covalency, for high-spin materials spin polarization may cause significant effects on both measured spins and on form factors, and orbital effects introduced via spin orbit coupling must be taken into account in discussing the form factors of singlet ions.

The LHFI was first observed by ESR [for $^{35}$Cl and Ir$^{4+}$ [$d^5$ low spin] doped in $K_2PtCl_6$ (24)] in 1953 and the determination of covalency parameters from such data was immediately realized (although subsequent discussions [e.g., Ref. (25)] indicate the possible ambiguities in this technique also). The first discussion of the determination of covalency parameters from magnetic neutron scattering experiments was not given until 1965 (26). This latecoming was the result of the lack of general availability of neutron sources, and their relative weakness which made very time-consuming the accurate measurement which is needed to detect, and to measure with any precision, the fairly small differences between observed magnetic scattering intensities and those predicted in the absence of covalency. The single crystal measurement of the reduced spin and the expanded form factor in NiO (27) gave the first reliable indication of deviations from ionic behavior, which led to the *Hubbard* and *Marshall* interpretation (26) of covalency effects in magnetic neutron scattering using the molecular orbital model. Apart from BaTbO₃ (28), almost all the compounds discussed below and from which information on covalency has been obtained, had previously been investigated by neutrons to determine the magnetic structure. Since the first discussion by *Hubbard* and *Marshall* (26) the state of knowledge has been reviewed only once (29), except briefly in a more wide ranging discussion (4). Considerable information has since been acquired making another discussion timely.

The MO model is discussed first and then the necessary elements of neutron scattering theory, with a brief description of experimental techniques. Finally there is a discussion of the data so far obtained.

## 2. The Molecular Orbital Bonding Model

### 2.1 The Simple MO Description of Octahedral Complexes

We discuss the situation pertaining to a simple octahedral complex of a $3d$ transition metal ion (e.g., $MF_6^{n-}$ or $MO_6^{m-}$). The details have been given many times previously [e.g., Refs. (11, 26, 30, 31)]. Bonding is assumed to place between the metal ion and the ligand ions. The relation of the MO formalism to the configuration interaction description has been discussed by *Owen* and *Thornley* (11).

The ionic wave functions of particular interest are the metal $3d$ orbitals and the ligand $2s$ and $2p$ orbitals. $\sigma$ bonds may be formed between the $3d_{x^2-y^2}$ and $3d_{z^2}$ ($e_g$) metal orbitals and the six ligand $p\sigma$ orbitals, and $\pi$ bonds between the $3d_{xy}$, $3d_{xz}$ and $3d_{yz}$ ($t_{2g}$) metal orbitals and the 12 ligand $p\pi$ orbitals. The filled inner shell orbitals may be neglected to first order in discussing magnetic neutron scattering, although not necessarily in other magnetic effects such as hyperfine interactions (32). The overlap of the $3d$ orbitals with $2p\sigma$ and $2p\pi$ orbitals is shown schematically in Fig. 1 and 2.

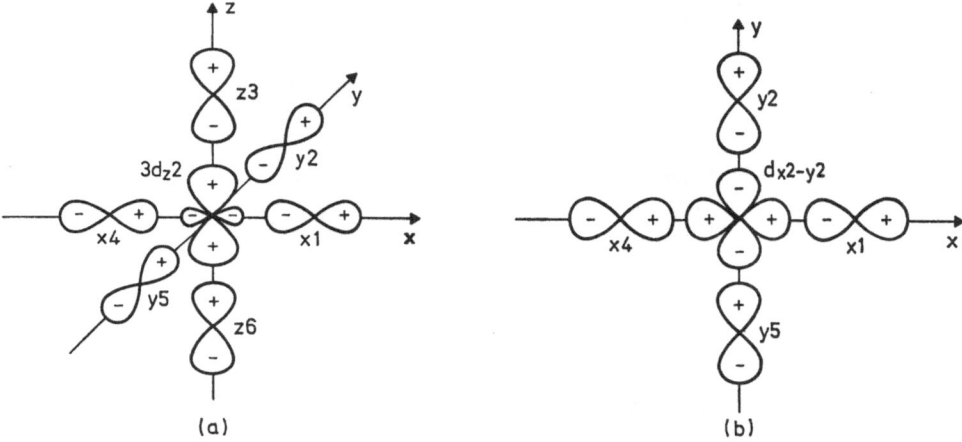

Fig. 1. Schematic overlap of $3d_{z^2}$ and $3d_{x^2-y^2}$ metal orbitals with $2p\sigma$ ligand orbitals. The orientation and numbering corresponds to that given in Eqs. (2.6) and (2.7). (Ligands 1, 2, 3 are located on the positive $x$, $y$, $z$ axes and 4, 5, 6 on the negative $x$, $y$, $z$ axes respectively)

The filled bonding orbitals of mainly ligand character are:

$$\psi_s^B = N_s^B \left( \chi_{2s} + \gamma_s d_\sigma + \gamma_{s\sigma} \chi_{2p\sigma} \right) \tag{2.1}$$

$$\psi_\sigma^B = N_\sigma^B \left( \chi_{2p\sigma} + \gamma_\sigma d_\sigma + \gamma_{\sigma s} \chi_{2s} \right) \tag{2.2}$$

$$\psi_\pi^B = N_\pi^B \left( \chi_{2p\pi} + \gamma_\pi d_\pi \right) . \tag{2.3}$$

7

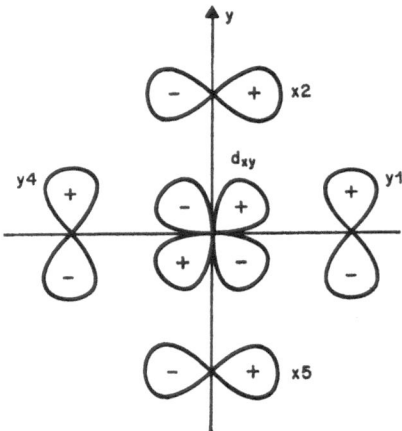

Fig. 2. Schematic overlap of one of the metal $t_{2g}$ orbitals ($3d_{xy}$) with ligand $2p\pi$ orbitals. The orientation and numbering corresponds to that given in Eq. (2.8)

$N_s^B$, $N_\sigma^B$ and $N_\pi^B$ are normalization constants and $\chi_{2s}$, $\chi_{2p\sigma}$ and $\chi_{2p\pi}$ are the appropriate linear combinations of ligand $2s$, $2p\sigma$ and $2p\pi$ orbitals, respectively. $\gamma_s$ and $\gamma_\sigma$ are the admixture parameters describing the $\sigma$ covalency between $d_\sigma$ and the ligand $2s$ and $2p\sigma$ orbitals respectively, and $\gamma_\pi$ measures the $\pi$ covalency between $d_\pi$ and $2p\pi$ orbitals.

Because the bonding orbitals are filled they do not directly contribute to the magnetic properties; the magnetic interactions associated with transition metal ions reflect the properties of the unpaired electrons in the antibonding orbitals, which have mainly $d$ character. Orthogonal to the bonding orbitals, they may be written:

$$\psi_\sigma = N_\sigma \left(d_\sigma - \lambda_\sigma \chi_{2p\sigma} - \lambda_s \chi_{2s}\right) \tag{2.4}$$

$$\psi_\pi = N_\pi \left(d_\pi - \lambda_\pi \chi_{2p\pi}\right) \tag{2.5}$$

In full, (2.4) and (2.5) are

$$\psi_{z2} = N_\sigma \left[d_z^2 - \frac{1}{\sqrt{12}}\, \lambda_\sigma \left(- 2p_{z3} + 2p_{z6} + p_{x1} - p_{x4} + p_{y2} - p_{y5}\right)\right. $$
$$\left. - \frac{1}{\sqrt{12}}\, \lambda_s \left(2s_3 + 2s_6 - s_1 - s_2 - s_4 - s_5\right)\right] \tag{2.6}$$

$$\psi_{x2-y2} = N_\sigma \left[d_{x2-y2} - \frac{1}{2}\, \lambda_\sigma \left(p_{x4} - p_{x1} + p_{y2} - p_{y5}\right)\right. $$
$$\left. - \frac{1}{2}\, \lambda_s \left(s_1 + s_4 - s_2 - s_5\right)\right] \tag{2.7}$$

$$\psi_{xy} = N_\pi \left[ d_{xy} - \frac{1}{2} \lambda_\pi \left( p_{y1} - p_{y4} + p_{x2} - p_{x5} \right) \right] \qquad (2.8)$$

$$\psi_{xz} = N_\pi \left[ d_{xz} - \frac{1}{2} \lambda_\pi \left( p_{z2} - p_{z5} + p_{y3} - p_{y6} \right) \right] \qquad (2.9)$$

$$\psi_{yz} = N_\pi \left[ d_{yz} - \frac{1}{2} \lambda_\pi \left( p_{x3} - p_{x6} - p_{z1} - p_{z4} \right) \right] \qquad (2.10)$$

The normalization constants in (2.6) to (2.10) are defined by $\langle \psi | \psi \rangle = 1$ and

$$N_\sigma = (1 - 2\lambda_\sigma S_\sigma - 2\lambda_s S_s + \lambda_\sigma^2 + \lambda_s^2)^{-\frac{1}{2}} \qquad (2.11)$$

$$N_\pi = (1 - 2\lambda_\pi S_\pi + \lambda_\pi^2)^{-\frac{1}{2}} \qquad (2.12)$$

where the overlap integrals $S$ are defined[3] by $\langle d | \chi \rangle$ and

$$S_\sigma = 2\langle d_\sigma | p_{2p\sigma} \rangle, \quad S_s = 2\langle d_\sigma | s_{2s} \rangle, \quad S_\pi = 2\langle d_\pi | p_{2p\pi} \rangle . \qquad (2.13)$$

Because of the orthogonality relations

$$\langle \psi_\sigma | \psi_\sigma^B \rangle = \langle \psi_\sigma | \psi_s^B \rangle = \langle \psi_\pi | \psi_\pi^B \rangle = 0 \qquad (2.14)$$

it follows that, to first order in $\lambda$, $\gamma$ and $S$

$$\lambda_\sigma = \gamma_\sigma + S_\sigma \qquad (2.15)$$

$$\lambda_s = \gamma_s + S_s \qquad (2.16)$$

$$\lambda_\pi = \gamma_\pi + S_\pi. \qquad (2.17)$$

Thus, we see that experiments such as magnetic neutron diffraction and spin resonance which measure $\lambda$ are sensitive to a combination of the covalency admixture parameter $\gamma$ and the overlap integral. However, it has become customary to treat $\lambda$ as a measure of the covalency. Clearly this situation is not entirely satisfactory, but may be tolerated in interpretive situations given the other approximations of the model.

Bonding also takes place between the ligand valence orbitals and the outer, initially unoccupied, $4s$ and $4p$ orbitals of the transition metal ion. Although it is generally thought that the metal $3d$ − ligand $2p$ bonding is the most significant covalent interaction in relatively ionic complexes, the $4s$ orbitals, being of larger radial extent, have a greater overlap with the ligand orbitals. Mattheiss, for

---

[3] Overlap integrals are often written as $S = \langle d | p \rangle$. In this case $N_\pi = (1 - 4\lambda_\pi S_\pi + \lambda_\pi^2)^{-\frac{1}{2}}$ etc.

example, has suggested (*33*) on the basis of APW band calculations that the charge redistribution in the transition metal monoxides may occur to as great an extent via mixing between metal $4s$ and $4p$ orbitals with ligand $2s$ and $2p$ orbitals as between the metal $3d$ — ligand $2p$ interaction. Such effects will clearly vary with the situation considered. For early transition metal ions where the $4s$ and $4p$ orbitals have radial distribution maxima of similar extent to the metal-ligand distance, charge transfer from ligand to metal will not cause any significant rearrangement of the charge distribution, but for compounds of the post-transition elements such as ZnO and ZnS, which, having tetrahedral coordination, are considered to be significantly covalent (*34*), the entire interaction must be via the outer orbitals. However, although charge transfer into the outer orbitals is interesting in the discussion of bonding and effective charges, for example, there is no involvement of unpaired electrons. Thus, to first order magnetic neutron diffraction, and spin resonance measurement of LHFI will not detect these interactions. This is not the case for other techniques such as nuclear quadrupole resonance (see below) and the Mössbauer effect isomer shift (*4*). To second order, where exchange effects and radial polarization are allowed for, magnetic moment densities may be induced in metal $4s$ orbitals and in nominally empty metal $3d$ orbitals. For the $d^5$ ions $Cr^+$ and $Mn^{2+}$ polarization of covalently occupied $4s$ orbitals is thought to be important in determining hyperfine interactions (*35*) and may also affect covalency parameters observed by neutron diffraction and spin resonance (*36*). For $d^3$ ions such as $Cr^{3+}$ and $Mn^{4+}$, effects attributed to spin polarization of charge transferred to the empty $e_g$ orbitals have been observed by neutron diffraction and spin resonance (see below).

A schematic molecular orbital diagram is shown in Fig. 3 for bonding involving ligand $2s$ and $2p$ orbitals and metal $3d$, $4s$ and $4p$ orbitals. The actual ordering of energy levels may be quite different in any real situation and polarization effects are not included. For comparison, a recent energy level calculation (*37*) for the $FeO_6^{9-}$ cluster (which has been investigated by both neutrons and spin resonance) is shown in Fig. 4. In Fig. 3 the convention has been followed that the $\sigma$ and $\pi$ antibonding orbitals, which contain the unpaired electrons in transition metal complexes, are less tightly bound than the mainly ligand bonding orbitals. This is probably a reasonable assumption in most cases but we should note that there is some experimental evidence from ultraviolet photoelectron spectroscopy (UVPES) and X-ray photoelectron spectroscopy (XRS) that this ordering scheme may not always be appropriate either for some $3d$ complexes (*38*) or for some rare earth ($4f$) compounds (*39*). In Fig. 4, in fact, we see that spin polarization effects place filled (spin up and spin down) $t_{1g}$ and $t_{1u}$ orbitals between the occupied (spin up) $t_{2g}$ and the empty (spin down) $t_{2g}$ orbitals, and the filled $e_g$ (spin up) orbitals also below the empty $t_{2g}$ orbitals.

The ligand field splitting $\Delta$ is shown in Fig. 3. From the simple MO model this is given by

$$\Delta = (\lambda_\sigma^2 - \lambda_\pi^2) (E_d - E_{2p}) + \lambda_s^2 (E_d - E_{2s}) \tag{2.18}$$

where $E_d$, $E_{2p}$ and $E_{2s}$ are the initial energies of the $3d$, $2p$ and $2s$ orbitals. Although (2.18) is a considerable simplification it is sufficient to make it clear that

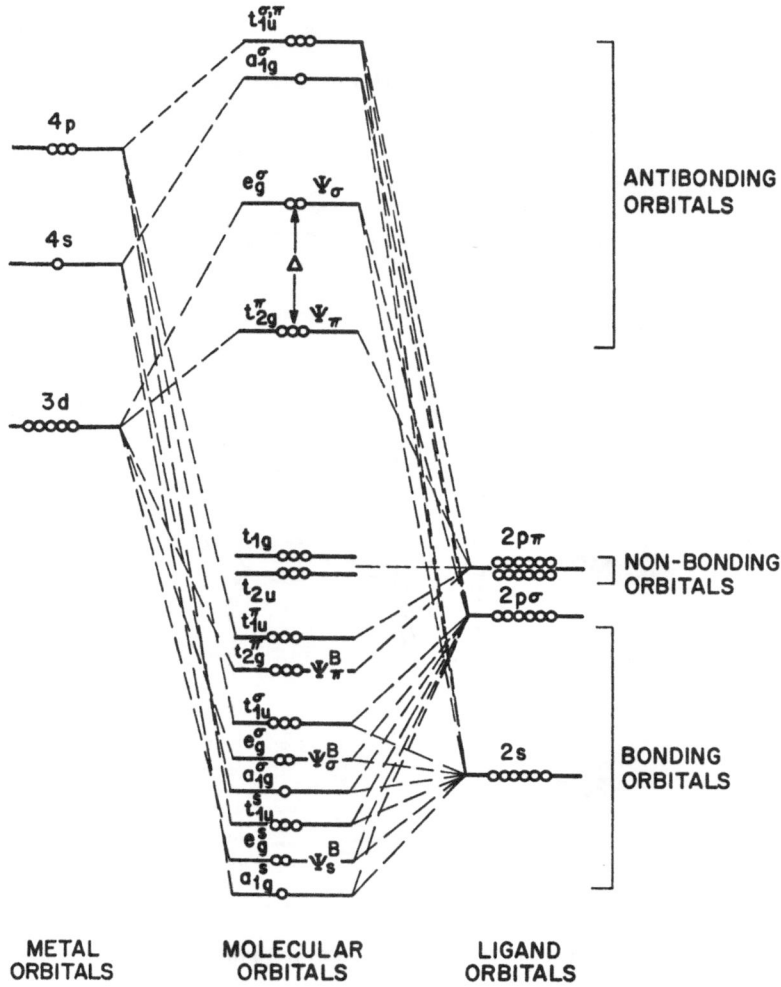

Fig. 3. Schematic molecular orbital diagram for bonding involving ligand $2s$ and $2p$ orbitals and metal $3d$, $4s$ and $4p$ orbitals. The bonding and antibonding orbitals described in Eqs. (2.1)—(2.5) are indicated and also the ligand field splitting $\Delta$

conclusions based on a simple crystal field splitting of the $3d$ levels, with no change in total energy, are of little significance. Site preference energies, Jahn-Teller splittings, and so forth are determined principally by the energies of the filled, mainly ligand, bonding orbitals. This is most clearly demonstrated of course when the properties of partially covalent $d^0$ complexes such as those of $Ti^{4+}$, $Nb^{5+}$ and $W^{6+}$ are discussed. The calculation of $\Delta$, and of the magnetic moment distributions determined by neutron scattering and LHFI measurement remains a principal target of first principles calculations on simple transition metal complexes.

11

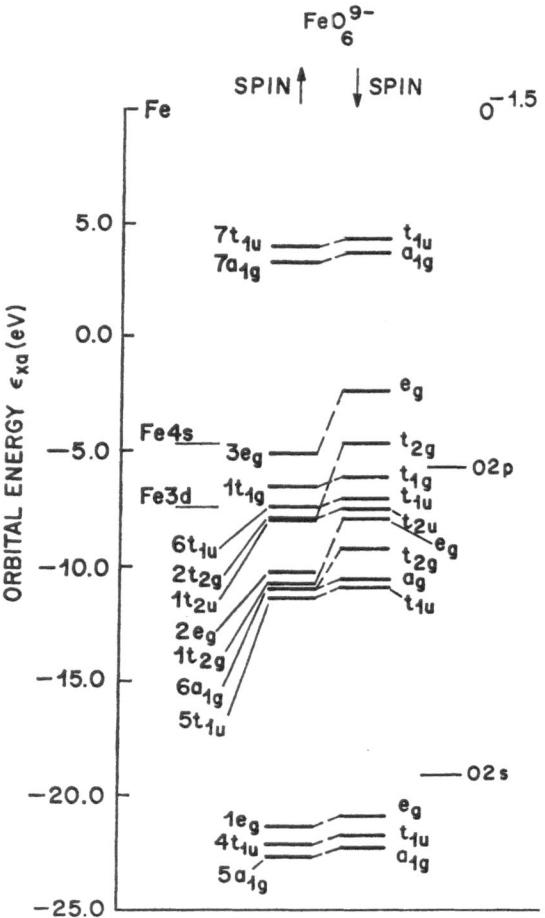

Fig. 4. Spin polarized energy levels of the $FeO_6^{9-}$ cluster from the multiple scattering calculation of Ref. (*37*)

## 2.2 The Effect of Covalency on the Magnetic Moment Distribution

In the ensuing discussion we assume a spin-only situation applies. The effect of orbital moments on the parameters observed in magnetic scattering experiments is complicated, although tractable (*1*). No doubt, more information on ions with orbitally degenerate ground states will be forthcoming, but to the present time, the great majority of experiments which have provided reliable information on bonding have been concerned with spin-only systems (principally octahedral $d^3$, $d^5$ and $d^8$ ions) where, for the most part, orbital effects, introduced by spin-orbit coupling, can be accounted for where necessary in a simple manner.

Consider a $\sigma$ antibonding orbital, $\psi_\sigma$, containing one electron. The magnetic moment distribution, $D(r)$, instead of being given simply by $(d_\sigma)^2$ as for the free ion is now given by $(\psi_\sigma)^2$. To second order:

$$D(r) = |\psi_\sigma(r)|^2 = d_\sigma^2 (1 + 2\lambda_\sigma S_\sigma + 2\lambda_s S_s - \lambda_\sigma^2 - \lambda_s^2)$$
$$- 2(\lambda_\sigma d_\sigma \chi_{2p\sigma} + \lambda_s d_\sigma \chi_{2s}) \tag{2.19}$$
$$+ (\lambda_\sigma^2 \chi_{2p\sigma}^2 + \lambda_\sigma^2 \chi_{2s}^2) + 2\lambda_\sigma \lambda_s \chi_{2s} \chi_{2p\sigma}$$

Similarly for $\psi_\pi$,

$$D(r) = |\psi_\pi(r)|^2 = d_\pi^2(1 + 2\lambda_\pi S_\pi - \lambda_\pi^2) - 2\lambda_\pi d_\pi \chi_{2p\pi} + \lambda_\pi^2 \chi_{2p\pi}^2 \tag{2.20}$$

We omit the explicit radial dependence of $d$ and $\chi$ (but note that the origin is the center of the metal wavefunction but $\chi$ is ligand centered) and assume that they are unchanged from the free ion situation. This assumption is not essential, however, as a neutron scattering experiment provides information on the radial distribution as well as the magnitude of the magnetic moment density.

Equations (2.19) and (2.20) show that $D(r)$ is composed of three terms — one of the form $d^2(1 + 2\lambda S - \lambda^2)$ associated with the metal ion, a term of the same sign which is associated with the ligand orbitals and an overlap term of opposite sign. In an isolated paramagnetic complex, such as is studied in the measurement of LHFI by paramagnetic resonance, the spin distribution is as given here, and the LHFI measures the spin density transferred to the ligands. For such experiments, it is customary to refer to the fraction of unpaired spin ($f$) transferred to a single ligand orbital when the metal $d$ orbitals of the appropriate symmetry are singly occupied. For two $e_g$ electrons (e.g., $d^5$ high spin, $d^8$) spin density $2\lambda_\sigma^2$ and $2\lambda_s^2$ is transferred to six $p\sigma$ and six $2s$ orbitals, respectively. For three $t_{2g}$ electrons (e.g., $d^3$, $d^5$ high spin) spin density $3\lambda_\pi^2$ is transferred to 12 $p\pi$ orbitals. Therefore

$$f_\sigma = \frac{1}{3}\lambda_\sigma^2, \ f_s = \frac{1}{3}\lambda_\sigma^2, \ f_\pi = \frac{1}{4}\lambda_\pi^2 \tag{2.21}$$

More accurately $f_\sigma = 1/3N_\sigma^2\lambda_\sigma^2$, etc., but (2.21) is correct to second order. The $f$'s have frequently also been used in discussing neutron diffraction data and are used as a measure of $\lambda$ via (2.21) but it should be noted that the relations of (2.21) are incorrect for situations where ligand orbitals are not singly occupied[4].

Although the discussion of the effects of covalency on the magnetic properties of transition metal ions reveals charge and spin density to be transferred from the metal ion to the ligands the net charge transfer is of course in the opposite direction. For each electron in a bonding orbital a quantity $(N^B)^2\gamma^2$ of charge is transferred from ligands to metal and for each electron in an antibonding orbital a quantity $N^2\lambda^2$ of charge is transferred from metal to ligands. There are 4

---

[4] One must beware of confusions in nomenclature. *Owen* and *Thornley* (*11*) use $\alpha$ and $\beta$ instead of $\lambda$ and $\gamma$. *Hubbard* and *Marshall* (*26*) used $A^2$ to denote $f$ and this convention has been followed in some neutron diffraction studies. Use of $A$ to denote fractional spin transfers however, can lead to confusion with the hyperfine interaction parameters determined experimentally by magnetic resonance.

electrons in the filled $\sigma$ bonding orbitals and 6 in the filled $\pi$ bonding orbitals. If there are $n$ $e_g$ and $m$ $t_{2g}$ antibonding electrons then neglecting $2s$ covalency, the net charge transfer $C$, towards the metal is

$$C = 4(N_\sigma^B)^2 \gamma_\sigma^2 + 6(N_\pi^B)^2 \gamma_\pi^2 - nN_\sigma^2 \lambda_\sigma^2 - mN_\pi^2 \lambda_\pi^2 \qquad (2.22)$$

If the approximation $\gamma = \lambda$ is used, then to second order

$$C = \lambda_\sigma^2 (4-n) + \lambda_\pi^2 (6-m) \qquad (2.23)$$

A relation of the type (2.23) was first used by *Owen et al.* (40) to rationalize the experimental covalency parameters of divalent metal ions (see below).

Eqs. (2.19) and (2.20) affect neutron scattering intensities via the magnetic form factor $f(\varkappa)$;

$$f(\varkappa) = \int e^{i\varkappa \cdot r} D(r) dr \qquad (2.24)$$

where $D(r)$ is the normalized spin density. $\varkappa$ is the scattering vector ($|\varkappa| = 4\pi \sin \theta/\lambda$ where $2\theta$ is the scattering angle and $\lambda$ the neutron wavelength). The form factor defines the angular dependence of the scattered intensity. By definition $f(0) = 1$. Neutron scattering by nuclei is isotropic (neglecting effects of thermal motion and absorption) because the small size of the nucleus relative to the neutron wavelength and the short-range neutron–nucleus interaction results in effective point scattering. This is in contrast to X-ray scattering by the charge distribution in an atom where the scattering interaction is of similar extent to the photon wavelength, and interference effects lead to a reduction of scattered intensity with increasing $\theta$ (defined by the scattering factor $f'(\varkappa) = \int e^{i\varkappa \cdot r} \varrho(r) dr$ where $\varrho(r)$ is the charge density distribution — in this case $f'(\varkappa)$ is generally normalized to the total charge on the atom, $Z$, *i.e.*, $f'(0) = Z$). For magnetic neutron scattering, occurring via interactions with unpaired valence electrons, the situation is analogous to X-ray scattering, except that the decrease of intensity with increasing scattering angle is generally somewhat steeper because only outer electrons, rather than the entire charge distribution, are involved.

It is convenient to rearrange (2.19)

$$D(r) = d_\sigma^2(1 - \lambda_\sigma^2 - \lambda_s^2) \qquad (2.25a)$$
$$+ 2(\lambda_\sigma S_\sigma d_\sigma^2 + \lambda_s S_s d_\sigma^2 - \lambda_\sigma d_\sigma \chi_{2p\sigma} - \lambda_s d_\sigma \chi_{2s}) \qquad (2.25b)$$
$$+ (\lambda_\sigma^2 \chi_{2p\sigma}^2 + \lambda_s^2 \chi_{2s}^2) + 2\lambda_\sigma \lambda_s \chi_{2s} \chi_{2p\sigma} \qquad (2.25c)$$

to give (a) a term in $d^2$ only, (b) an overlap term and (c) a term in $\chi^2$ only. At $\varkappa = 0$.

$$f_a(0) = 1 - \lambda_\sigma^2 - \lambda_s^2$$
$$f_b(0) = 0 \qquad (2.26)$$
$$f_c(0) = \lambda_\sigma^2 + \lambda_s^2 .$$

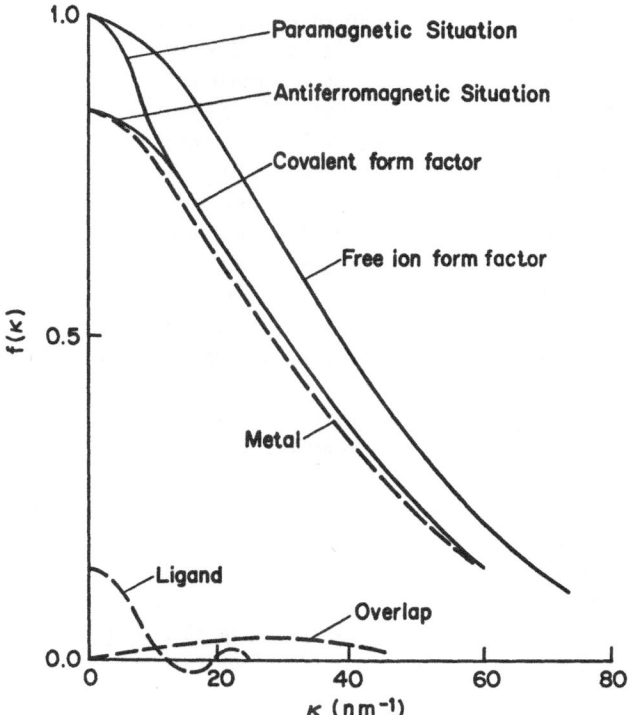

Fig. 5. Typical form factor for a $3d$ ion such as $Ni^{2+}$ based on the simple MO theory of Ref. (26). The free ion form factor and the covalent form factor for both paramagnetic and anti-ferromagnetic situations are shown. The three components of the covalent form factor, consisting of contributions from the metal ion spin, the ligand spin and the overlap spin are shown by broken lines

The effect on the whole form factor for an ion such as $Ni^{2+}$ is shown in Fig. 5. The metal only term (a) behaves as the free ion curve [given by putting $d(r)$ in (2.24)], but is reduced by the factor $1 - \lambda_\sigma^2 - \lambda_s^2$. The ligand term (c), involving the most extended spin density, has a sharply peaked form factor. For the spherically averaged situation, $f_c$ is given closely by

$$f_c(\varkappa) = \frac{\sin(\varkappa R)}{\varkappa R} \left[\lambda_\sigma^2 \int j_0(\varkappa r)p^2(r)dr + \lambda_s^2 \int j_0(\varkappa r)s^2(r)dr\right] \qquad (2.27)$$

where $j_0(\varkappa r)$ is the zero order spherical Bessel function and $p^2(r)$ and $s^2(r)$ are the charge densities of $2p$ and $2s$ orbitals respectively, both centered on the origin. The term $\sin(\varkappa R)/\varkappa R$ arises from the change of origin of the ligand orbitals ($R$ is the metal-ligand distance) and determines the narrow 'forward peak' of $f_c(\varkappa)$. Although this is an oscillatory function, $f_c(\varkappa)$ is only significant for $\varkappa \lesssim 1.0 - 1.5 Å^{-1}$. In this region the shape of the form factor is significantly changed from the free ion situation, and measurement of the forward peak should provide, in

15

principle, a fairly direct measurement of $\lambda$ as pointed out by *Hubbard* and *Marshall* (26). Owing to experimental difficulties this has yet to be realized — only two measurements of the effect have been reported so far (20, 41).

If the form factor is determined at $\varkappa \gtrsim 1.0 - 1.5\text{Å}^{-1}$ the effective moment will be reduced from the real moment by the factor $(1 - \lambda_\sigma^2 - \lambda_s^2)$, and this is the means by which most information on bonding has been obtained by neutron diffraction. By measuring the absolute intensity of the lowest angle (and generally most intense) magnetic reflection of a magnetically ordered material (frequently occurring at $\varkappa \approx 1.2 - 1.5\text{Å}^{-1}$) the covalency may be determined without knowledge of the real form factor (the difference from the calculated free ion form factor is very small in this region). This method is applicable to powders where accurate form factor determinations are generally not possible. Only by collecting full three-dimensional data on single crystals can the complete spin distribution be obtained by Fourier inversion methods to give detailed information on the form factor, and the contributions from the ligand, overlap and metal regions.

Although the discussion just given assumes the presence of the ligand moment, this is in fact completely quenched in most powders studied to date, which have all been antiferromagnets. In an antiferromagnet the net spin over the magnetic unit cell is zero and neighboring magnetic moments are generally oppositely aligned. This is illustrated in one dimension in Fig. 6. The spin induced at a ligand

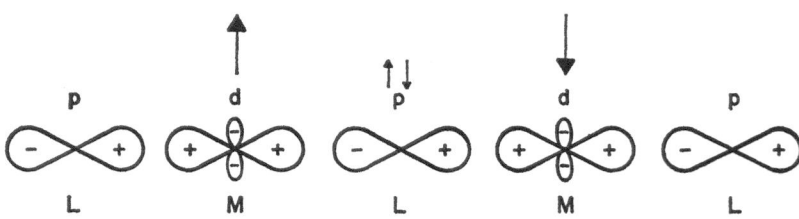

Fig. 6. Orbitals for a linear antiferromagnet showing a net spin of zero transferred to the ligands

by a metal ion with spin up is exactly cancelled by the contribution from the neighbor with spin down. The resulting form factor is shown in Fig. 5. The magnetic moment per magnetic ion is thus reduced by the factor $(1 - \lambda_\sigma^2 - \lambda_s^2)$. Eqs. (2.19) and (2.25) can thus be rewritten

$$D(\boldsymbol{r}) = d_\sigma^2(1 + 2\lambda_\sigma S_\sigma + 2\lambda_s S_s - \lambda_\sigma^2 - \lambda_s^2)$$
$$- 2(\lambda_\sigma S_\sigma \chi_{2p\sigma} + \lambda_s d_\sigma \chi_{2s}) \tag{2.28}$$

and

$$D(\boldsymbol{r}) = d_\sigma^2(1 - \lambda_\sigma^2 - \lambda_s^2)$$
$$+ 2(\lambda_\sigma S_\sigma d_\sigma^2 + \lambda_s S_s d_\sigma^2 - \lambda_\sigma d_\sigma \chi_{2p\sigma} - \lambda_s d_\sigma \chi_{2s}). \tag{2.29}$$

16

From (2.28) it is apparent that, because of the small value of $d_\sigma^2$ close to the ligand nuclei, in this region $D(\mathbf{r})$ is dominated by the second negative term. By symmetry, the moment density is zero at the ligand nucleus and thus each magnetic ion of positive moment is surrounded by a region of negative moment and vice versa[5]).

As shown in Fig. 5 the overlap form factor $f_b(\varkappa)$ is zero at $\varkappa = 0$ and passes through a maximum as $\varkappa$ increases. The simple molecular orbital model thus predicts a form factor for an antiferromagnet reduced from the free ion value, but somewhat expanded in shape because of the overlap moment. Although many other factors such as the introduction of orbital effects via spin orbit coupling, the polarization of filled inner shells, the polarization of partially occupied outer orbitals, and variations in the $3d$ radial functions need to be considered before a detailed form factor analysis can be attempted in any particular situation, it was satisfying that the first accurately determined ionic form factor, for $Ni^{2+}$ in NiO (27), behaved qualitatively in this manner. Unfortunately few other form factors have been accurately determined since then, but the behavior is not always so apparently straightforward.

The analysis for $\pi$ bonding is exactly analogous to that given for $\sigma$ bonding. The moment reductions expected for high spin ions in octahedral coordination are given in Table 1. Also given are the covalency parameters determined by LHFI measurement for the spin only situations $d^3$, $d^5$ and $d^8$ (11). For $d^3$ and $d^8$ ions neutron and LHFI data should be equivalent. For $d^5$ ions, LHFI gives the difference of $\lambda_\sigma$ and $\lambda_\pi$ whereas neutron diffraction measures the total covalency; comparison of the data from the two techniques, therefore, should give $\sigma$ and $\pi$ antibonding parameters independently.

It is interesting to compare the moment reductions of Table 1 with the information that may be obtained from ligand nuclear quadrupole coupling constants. The electric field gradient at the nucleus of an atom caused by a single electron is given by

$$eq = e\langle 3\cos^2\theta - 1\rangle \langle r^{-3}\rangle \qquad (2.30)$$

where $e$ is the electronic charge.

In an axially symmetric field gradient such as occurs for a ligand in an octahedral complex with the bond axis along $z$, $\langle 3\cos^2\theta - 1\rangle = 4/5$ for $p_z$ and $-2/5$ for $p_x$ and $p_y$ electrons. The quadrupole coupling constant ($QCC$) determined in nuclear quadrupole reasonance spectroscopy or from quadrupole interactions observed in magnetic resonance is given by $eqQ$ where $Q$ is the quadrupole moment of the ligand nucleus. For a filled ligand valence shell $q = 0$ and quadrupole coupling constants are small, arising from lattice, shielding and overlap effects. This is the situation for the alkali halides for example. Quadrupole interactions are also not observed if $I = 1/2$ or 0. Unfortunately $^{19}F$ has $I = 1/2$ and $^{16}O$ has $I = 0$.

---

[5]) This assumes the sign of $\lambda$ to be positive. This is so for the model described but is not necessarily so for complexes where ligand-to-metal back bonding might occur (e.g., in hexacyanide complexes). Such situations have not yet been investigated by neutron diffraction.

Table 1. Moment reductions for antiferromagnets. (Octahedral Complex)[a]

| $d$ Electrons | Ions | Moment Reduction | | Covalency Parameter from LHFI (Spin Only Ions) |
|---|---|---|---|---|
| 1 $(t_{2g}^1, e_g^0)$ | Ti$^{3+}$, V$^{4+}$ | $\lambda_\pi^2 = 4f_\pi$ | | — |
| 2 $(t_{2g}^2, e_g^0)$ | V$^{3+}$, Cr$^{4+}$ | $\lambda_\pi^2 = 4f_\pi$ | | — |
| 3 $(t_{2g}^3, e_g^0)$ | V$^{2+}$, Cr$^{3+}$, Mn$^{4+}$ | $\lambda_\pi^2 = 4f_\pi$ | | $\lambda_\pi^2$ $(f_\pi)$ |
| 4 $(t_{2g}^3, e_g^1)$ | Cr$^{2+}$, Mn$^{3+}$ | $\frac{1}{4}(\lambda_\sigma^2 + 3\lambda_\pi^2 + \lambda_s^2) = \frac{3}{4}(f_\sigma + 4f_\pi + f_s)$ | | — |
| 5 $(t_{2g}^3, e_g^2)$ | Cr$^+$, Mn$^{2+}$, Fe$^{3+}$ | $\frac{1}{5}(2\lambda_\sigma^2 + 3\lambda_\pi^2 + 2\lambda_s^2) = \frac{6}{5}(f_\sigma + 2f_\pi + f_s)$ | | $\lambda_s^2, \frac{1}{3}\lambda_\sigma^2 - \frac{1}{4}\lambda_\pi^2$ $(f_s, f_\sigma - f_\pi)$ |
| 6 $(t_{2g}^4, e_g^2)$ | Fe$^{2+}$, Co$^{3+}$ | $\frac{1}{2}(\lambda_\sigma^2 + \lambda_\pi^2 + \lambda_s^2) = \frac{1}{2}(3f_\sigma + 4f_\pi + 3f_s)$ | | — |
| 7 $(t_{2g}^5, e_g^2)$ | Co$^{2+}$, Ni$^{3+}$ | $\frac{1}{3}(2\lambda_\sigma^2 + \lambda_\pi^2 + 2\lambda_s^2) = \frac{2}{3}(3f_\sigma + 2f_\pi + 3f_s)$ | | — |
| 8 $(t_{2g}^6, e_g^2)$ | Ni$^{2+}$, Cu$^{3+}$ | $\lambda_\sigma^2 + \lambda_s^2 = 3(f_\sigma + f_s)$ | | $\lambda_s^2, \lambda_\sigma^2$ $(f_s, f_\sigma)$ |
| 9 $(t_{2g}^6, e_g^3)$ | Cu$^{2+}$ | $\lambda_\sigma^2 + \lambda_s^2 = 3(f_\sigma + f_s)$ | | — |

[a] $\lambda_\sigma^2 = 3f_\sigma$, $\lambda_\pi^2 = 4f_\pi$.

Although fluoride hosts are the most popular for studying LHFI, $QCC$ cannot be observed, but a small amount of data are now available for oxides doped with $^{17}$O (I $= 5/2$). Most information, however, has been obtained for the heavier halides - in particular for $^{35}$Cl and $^{37}$Cl (I $= 3/2$).

Because of the $\langle r^{-3} \rangle$ term ligand $QCC$ will be dominated by effects involving the outer $p$ orbitals. The net $QCC$ will reflect the difference in population of the $p_z$ orbitals (involved in $\sigma$ bonding) and the $p_x$ and $p_y$ orbitals (involved in $\pi$ bonding). Because of lattice contributions to $q$ and Sternheimer shielding and antishielding effects it is generally difficult to estimate or interpret $QCC$ with any confidence. For the halogens, however, experimental $QCC$ may be calibrated by reference to $Cl_2$ or Cl atoms where there is one hole in the $p_z$ orbital. It is convenient to define $f_Q$ (42),

$$f_Q = \frac{eqQ \text{ crystal}}{eqQ \text{ atom}} \qquad (2.31)$$

If outer $p$ orbitals only are considered $f_Q$ is related to the number of electrons in the $p_x$, $p_y$ and $p_z$ orbitals by (43)

$$f_Q = \frac{N_x + N_y}{2} - N_z \, . \qquad (2.32)$$

If it is assumed that bonding takes place with the metal $d$ orbitals only, $f_Q$ may be related to $\lambda_\sigma$ and $\lambda_\pi$. Details are given in Table 2 for an octahedral complex with each ligand bonded to one metal atom. Because $eqQ$ reflects the charge

Table 2. Relation of $f_Q$ to covalency parameters for an octahedral complex[a])

| $d$ Electrons | $f_Q$ | $d$ Electrons | $f_Q$ |
|---|---|---|---|
| $d^0$ | $\frac{2}{3}\lambda_\sigma^2 - \frac{1}{2}\lambda_\pi^2 = 2(f_\sigma - f_\pi)$ | $d^5$ (low spin) | $\frac{2}{3}\lambda_\sigma^2 - \frac{1}{12}\lambda_\pi^2 = 2f_\sigma - \frac{1}{3}f_\pi$ |
| $d^1$ | $\frac{2}{3}\lambda_\sigma^2 - \frac{5}{12}\lambda_\pi^2 = 2f_\sigma - \frac{5}{3}f_\pi$ | $d^6$ (high spin) | $\frac{1}{3}\lambda_\sigma^2 - \frac{1}{6}\lambda_\pi^2 = f_\sigma - \frac{2}{3}f_\pi$ |
| $d^2$ | $\frac{2}{3}\lambda_\sigma^2 - \frac{1}{3}\lambda_\pi^2 = 2f_\sigma - \frac{4}{3}f_\pi$ | $d^6$ (low spin) | $\frac{2}{3}\lambda_\sigma^2 = 2f_\sigma$ |
| $d^3$ | $\frac{2}{3}\lambda_\sigma^2 - \frac{1}{4}\lambda_\pi^2 = 2f_\sigma - f_\pi$ | $d^7$ (high spin) | $\frac{1}{3}\lambda_\sigma^2 - \frac{1}{12}\lambda_\pi^2 = f_\sigma - \frac{1}{3}f_\pi$ |
| $d^4$ (high spin) | $\frac{1}{2}\lambda_\sigma^2 - \frac{1}{4}\lambda_\pi^2 = \frac{3}{2}f_\sigma - f_\pi$ | $d^7$ (low spin) | $\frac{1}{2}\lambda_\sigma^2 = \frac{3}{2}f_\sigma$ |
| $d^4$ (low spin) | $\frac{2}{3}\lambda_\sigma^2 - \frac{1}{6}\lambda_\pi^2 = 2f_\sigma - \frac{2}{3}f_\pi$ | $d^8$ | $\frac{1}{3}\lambda_\sigma^2 = f_\sigma$ |
| $d^5$ (high spin) | $\frac{1}{3}\lambda_\sigma^2 - \frac{1}{4}\lambda_\pi^2 = f_\sigma - f_\pi$ | $d^9$ | $\frac{1}{6}\lambda_\sigma^2 = \frac{1}{2}f_\sigma$ |

a) $\lambda_\sigma^2 = 3f_\sigma$, $\lambda_\pi^2 = 4f_\pi$.

distribution in the ligand orbitals it is determined by the net effect of charge transfer in the bonding and antibonding orbitals (the values in Table 2 assume $\gamma = \lambda$). Thus $f_Q$ is not zero for diamagnetic $d^0$ or low spin $d^6$ complexes. Generally $f_Q$ and LHFI do not provide the same combination of $\lambda_\pi$ and $\lambda_\sigma$, but they do for $d^5$ high spin complexes. Comparison of $QCC$ data and data from LHFI and neutron diffraction should be valuable as a check on the data obtained by each and might allow an assessment of $QCC$ data for other nonmagnetic systems such as the A and B metals. A discussion of $f_Q$ for various metal-ligand combinations is given in Ref. (4). Differences between the covalency parameters determined might also provide an indication of the importance of bonding with outer metal $s$ and $p$ orbitals. $\sigma$ bonding with $s$ orbitals and $\sigma$ and $\pi$ bonding with $p$ orbitals will both affect $f_Q$. Few comparisons are possible at the moment because magnetic data has mainly been concerned with oxide and fluoride complexes. For Mn$^{2+}$ however, we might note that $f_\sigma - f_\pi$ has been determined as 0.0% and 0.6% for the two Cl atoms in CsMnCl$_3$ (44) and 4.3% for Mn$^{2+}$ in K$_4$CdCl$_6$ (45) whereas $f_Q$ was 9.4% and 16.7% for the former compound (44) and 18.7% for KMnCl$_3$ (46). Although because of the approximations made, $f_Q$ data should be interpreted with some caution, and correction should also be made for the fact that the chlorines are bonded to more than one metal in a nonlinear situation, this comparison may indicate a significant contribution to $f_Q$ from $\sigma$ bonding to the $Mn^{2+}$ $4s$ and $4p$ orbitals. In tetrahedral $d^{10}$ complexes at least, such bonding is obviously important and is reflected in experimental $f_Q$; for $ZnCl_4{}^{2-}$ for example $f_Q = 16.2\%$ (47). More LHFI and especially more neutron data are needed for complexes of the heavier halides and other ligands than $O^{2-}$ and $F^-$.

# 3. Neutron Scattering, Theory and Techniques

The theory of the magnetic scattering of thermal neutrons has been described in some detail by *Marshall* and *Lovesey* (*1*). Only a brief outline of those relations relevant to the present discussion will be given here. This limits the discussion to elastic scattering by spin-only systems[6]. Experimental techniques such as time-of-flight diffractometery are not discussed as no applications have yet been made to the subject under discussion.

## 3.1 Coherent and Incoherent Nuclear Scattering

The incident and diffracted neutron wave-vectors are denoted by $k$ and $k'$, respectively ($k = 2\pi/\lambda$) and the scattering vector $\varkappa = k - k'$. The neutron state $|k\rangle$ can be represented by a plane wave $\exp(i k \cdot r)$.

The cross-section of a target ($\sigma$) is given by

$$\sigma = \frac{\text{Interactions per second}}{\text{Neutron flux per cm}^2 \text{ per second}} \ (\text{cm}^2) \tag{3.1}$$

The total cross section, $\sigma_T$ is the sum of the absorption ($\sigma_a$) and scattering cross sections ($\sigma_s$). $\sigma_a$ is small for most elements in the case of low-energy thermal neutrons $\left(\text{few meV energy, } E = \dfrac{\hbar^2 k^2}{2m_n}\right)$. Where it is large, neutron diffraction is not practical [but the $^{10}\text{B}(n,\alpha)^7\text{Li}$ reaction is used for neutron counting and Cd is a convenient shielding material to define sample heights]. Absorption corrections are angular dependent and are necessary in form factor determinations. It is simplest to measure the transmission experimentally to determine the linear absorption coefficient, from which standard corrections may be estimated. The correction is generally small — *e.g.*, for polycrystalline MnO (*48*) less than one percent correction was necessary for intensities between 0° and 90° 2θ, although $\sigma_a$ for Mn (7.6 barns[7]) is larger than for most elements. The small absorption correction is one advantage possessed by neutron diffraction relative to X-ray scattering.

For nuclear scattering, which is short range and isotropic and may be formalized within the Born approximation, the cross section of a single nucleus is related to the scattering length $b$ (which may be positive or negative)

$$\sigma = 4\pi |b|^2. \tag{3.2}$$

---

[6] A form factor may in principle be determined by inelastic magnetic scattering also, but no data on bonding is available from this type of experiment (*47a*).

[7] 1 barn = $10^{-28}$ m².

The scattering from a sample is not generally isotropic, however, and it is convenient to define a differential cross section

$$\frac{d^2\sigma}{d\Omega dE} \qquad (3.3)$$

(barns per steradian per unit energy transfer).

For elastic scattering this is replaced by $d\sigma/d\Omega$.

The master formula for the scattering of neutrons of initial and final states $|k\rangle$ and $|k'\rangle$ by a sample of initial and final states $|\lambda\rangle$ and $|\lambda'\rangle$ is (1)

$$\frac{d^2\sigma}{d\Omega dE} = \frac{k'}{k}\left(\frac{m}{2\pi\hbar^2}\right)^2 \sum_{\lambda,s} p_\lambda p_s \sum_{\lambda',s'} |\langle k',s',\lambda'|\hat{V}|k,s,\lambda\rangle|^2 \delta\left(\frac{\hbar^2}{2m_n}(k'^2-k^2)+E_{\lambda'}-E_\lambda\right) \qquad (3.4)$$

where $s$ and $s'$ refer to the initial and final spin states of the neutron and the $\delta$-function describes energy conservation in the case of inelastic scattering ($m_n$ is the neutron mass). $\hat{V}$ is the interaction potential for scattering and the cross section is summed over all possible final and initial states of $s$ and $\lambda$. $p_s$ and $p_\lambda$ are the probabilities of the initial states. $p_s = 1/2$ for $s = \pm 1/2$ for unpolarized neutrons; in this case magnetic and nuclear cross sections add and there is no interaction between them.

For elastic nuclear scattering from an assembly of $N$ atoms of one element with position vector $\boldsymbol{R}$

$$\frac{d\sigma}{d\Omega} = \sum_{ij} e^{i\varkappa \cdot (\boldsymbol{R_i} - \boldsymbol{R_j})} \overline{b_i^* b_j} . \qquad (3.5)$$

The scattering lengths will depend on the particular isotope at each site and on the nuclear spin orientations. If $i$ and $j$ refer to different sites there is no correlation and

$$\overline{b_i^* b_j} = |\bar{b}|^2 .$$

For $i=j$

$$\overline{b_i^* b_j} = \overline{|b|^2} . \qquad (3.6)$$

Generally

$$\overline{b_i^* b_j} = |\bar{b}|^2 + \delta_{i,j}\left(\overline{|b|^2} - |\bar{b}|^2\right)$$

Thus $d\sigma/d\Omega$ is composed of two terms: an incoherent term which is isotropic and a coherent term which is not.

$$\frac{d\sigma}{d\Omega} = \left(\frac{d\sigma}{d\Omega}\right)_{\text{coh}} + \left(\frac{d\sigma}{d\Omega}\right)_{\text{incoh}} \qquad (3.7)$$

where

$$\left(\frac{d\sigma}{d\Omega}\right)_{\text{coh}} = |\bar{b}|^2 \left| \sum_i e^{i\varkappa \cdot R_i} \right|^2 \tag{3.8}$$

and

$$\left(\frac{d\sigma}{d\Omega}\right)_{\text{incoh}} = N \left[ \overline{|b|^2} - |\bar{b}|^2 \right] = N \overline{|b - \bar{b}|^2} \tag{3.9}$$

Coherent scattering lengths and incoherent cross sections have been tabulated [e.g., Ref. (2), Ref. (49)]. Coherent scattering lengths vary somewhat randomly from element to element (and for isotopes of a given element). Two advantages that neutron scattering has in comparison to X-ray scattering is that adjacent elements in the periodic table are generally easily distinguished and that light elements (e.g., D, C, N, O, F) are frequently as strong scatterers as heavy ones. The absence of an angular dependence of $\bar{b}$ is also an advantage in collecting data at high angles.

For a system of different atoms randomly distributed on a lattice, the incoherent cross-section arises from the incoherent scattering of the individual elements and from the disorder scattering due to the random distribution of the different atoms. This latter cross-section is

$$\left(\frac{d\sigma}{d\Omega}\right)_{\text{incoh}} = \sum_a c_a |\bar{b}_a|^2 - \left| \sum_a c_a \bar{b}_a \right|^2 \tag{3.10}$$

which for two elements of concentration $c_1$ and $c_2$ ($c_1 + c_2 = 1$) becomes

$$\left(\frac{d\sigma}{d\Omega}\right)_{\text{incoh}} = c_1 c_2 (\bar{b}_1 - \bar{b}_2)^2 . \tag{3.11}$$

This term disappears for $\bar{b}_1 = \bar{b}_2$. The coherent scattering length is simply the average $\sum\limits_a c_a \bar{b}_a$.

The scattering length of any nucleus depends on the orientation of the nuclear spin $I$ to the neutron spin. The interacting system can have total spin $I + 1/2$ or $I - 1/2$ giving different scattering lengths $b_+$ and $b_-$ in the ratio

$$\frac{b_+}{b_-} = \frac{I+1}{2I+1} \bigg/ \frac{I}{2I+1} . \tag{3.12}$$

Thus

$$b_{\text{coh}} = \left(\frac{I+1}{2I+1}\right) b_+ + \left(\frac{I}{2I+1}\right) b_- \tag{3.13}$$

and

$$b_{\text{incoh}} = \frac{I(I + 1)}{(2I + 1)^2} (b_+ - b_-)^2 . \qquad (3.14)$$

If $b_+$ and $b_-$ are of different sign (*e.g.*, H, V) there is a small coherent cross section and a large incoherent cross section. For this reason vanadium is used as a calibrant in incoherent scattering cross-section measurements (see below) and as a sample container (for polycrystalline materials) for many of the experiments described below so that unwanted peaks are not introduced into the diffraction pattern. On the other hand, if an element has one isotope of zero nuclear spin in large abundance the scattering is almost entirely coherent (*e.g.*, O, Fe).

## 3.2 Nuclear Bragg Scattering

For scattering from a crystal, 3.8 reduces to

$$\left(\frac{d\sigma}{d\Omega}\right)_{\text{coh}} = \frac{N(2\pi)^3}{v_0} \sum_{\tau} \delta(\varkappa - \tau)|F_N(\varkappa)|^2 \qquad (3.15)$$

where $v_0$ is the volume of the unit cell[8]), $N$ is the number of unit cells in the sample and $\tau$ is a reciprocal lattice vector ($|\tau_{hkl}| = 2\pi/d_{hkl}$). The $\delta$-function defines the Bragg condition. $F_N(\varkappa)$ is the nuclear unit cell structure factor:

$$F_N(\varkappa) = \sum_n \bar{b}_n e^{i\varkappa \cdot r_n} e^{-W_n(\varkappa)} = \sum_n \bar{b}_n e^{2\pi i(hu_n + kv_n + lw_n)} e^{-W_n(\varkappa)} \qquad (3.16)$$

where the sum is over the $n$ atoms in the unit cell, with coordinates $r_n$ ($= u_n a + v_n b + w_n c$) and $h$, $k$ and $l$ are Miller indices. $a$, $b$ and $c$ are the direct lattice vectors and $u_n$, $v_n$, and $w_n$ fractional coordinates. The Debye-Waller factor $\exp[-W_n(\varkappa)]$ arises from atomic vibrations and is both angular and temperature dependent. For many of the powder experiments described below where comparison was made between the intensities of a small number of low angle reflections it was sufficient to estimate an overall Debye-Waller correction ($\leq 1\%$ at 4.2K) from the zero-point formula

$$2W = \frac{5740}{\bar{A}\Theta} \frac{\sin^2\theta}{\lambda^2} \qquad (3.17)$$

where $\Theta$ is the Debye temperature of the sample and $A$ the average atomic weight. Where more complete data are collected it is necessary to determine individual

---

[8]) The first term derives from the sum over the unit cells: $\left|\sum_l e^{i\varkappa \cdot l}\right|^2 = \frac{N(2\pi)^3}{v_0}$

$\sum_{\tau} \delta(\varkappa - \tau)$ where the $l$ vectors denote the positions of the unit cells.

atom Debye-Waller factors from the data. $W(\varkappa)$ may be expressed (1) as a power series in $(\varkappa \cdot \boldsymbol{u})$, which in the harmonic approximation reduces to

$$W_n(\varkappa) = \frac{1}{2} \langle (\varkappa \cdot \boldsymbol{u}_n)^2 \rangle = 8\pi^2 \, \overline{u_n^2} \, \frac{\sin^2\theta}{\lambda^2} \qquad (3.18)$$

$8\pi^2 \, \overline{u_n^2} = B_n$ is the angular independent temperature factor generally quoted. $\overline{u_n^2}$ is the mean square atomic displacement from the ideal lattice position. The calculation of temperature factors from lattice dynamical models has been discussed by *Hewat* (50). At low temperatures, temperature factors are small, arising from zero-point motion effects only, and it is often acceptable in nuclear structure refinements to use an overall value. This approximation is not always satisfactory where very accurate magnetic intensities must be measured. In the study of poly-cristalline MnO (48) and US (19) it was helpful to use calculated temperature factors. At low temperatures $B_n$ is proportional to $m_n^{-1/2}$ where $m$ is the atomic weight. At higher temperatures for simple (e.g. rocksalt) structures where the coordination numbers of anions and cations are the same, the temperature factors for all atoms in the cell should become the same if a harmonic model with dominant nearest neighbor forces is assumed. In many materials this will not be observed in practice owing to anharmonic effects, data inaccuracy, effects of non-stoichiometry and so forth. It has not been necessary to introduce anisotropic temperature factors in most cases discussed in this review.

In a single crystal experiment a monochromatic beam (produced by Laue reflection from a single crystal monochromator) is used, and the crystal is free to rotate. The observed intensities are not $\delta$-functions owing to the finite crystal size, the mosaic spread in the crystal and collimation and wavelength spreads in the incident beam, and the measured intensity is integrated over a rocking curve. In this situation the integrated intensity over the rocking curve is [from (3.15)]

$$I = I_0 V Q^{2\theta} \qquad (3.19)$$

where $I_0$ is the intensity of the incident beam at the wavelength $\lambda$ used, $V$ the volume of the crystal[9]) exposed to the beam and

$$Q^{2\theta} = \frac{\lambda^3 |F_N(\varkappa)|^2}{v_0^2 \sin 2\theta} . \qquad (3.20)$$

The techniques of single-crystal diffractometry have been discussed by *Arndt* and *Willis* (51). We should note that extinction is a very serious problem in the determination of accurate magnetic intensity data from single crystals. Although extinction must always be accounted for in conventional crystallographic studies, it is particularly important to make proper correction in polarized neutron experiments where the ratio of magnetic to nuclear structure factors is determined.

---

[9]) Equation (3.19) is of course only correct in the small crystal limit where extinction effects and other beam attenuation processes are small.

If not, incorrect conclusions about magnetic moment reductions and the shapes of form factors may be made. Discussion of the extinction corrections (52) in form factor determinations has recently been given in the case of Tb(OH)$_3$ (53) where the intensity of some reflections was reduced by as much as 90%, and K$_2$NaCrF$_6$ (22). Because of the small crystal size, extinction has not been observed to be significant in any work with polycrystalline samples, which is one of the principal advantages of the latter technique. Preferred orientation can be a nuisance in powder work (especially with X-rays) but does not appear to have been significant in the experiments discussed below.

Modern powder techniques (see below) allow the collection of data of high accuracy out to high scattering angles. A single crystal study enables the collection of more data because overlapping reflections are separated. This is useful in the study of some high symmetry (especially cubic) systems where there can be a serious problem in powder work, or where a large number of parameters must be determined. Single crystal studies have been necessary also in almost all form factor determinations. However, the absence of extinction effects, and the often considerably easier preparation of polycrystalline samples, together with the more rapid data collection make powder work very attractive in many circumstances. As mentioned above, most of the covalency parameters determined by neutron scattering have, in fact, been obtained using polycrystalline samples. In general these were of sufficiently high symmetry so that accurate comparison of low-angle nuclear and magnetic intensities could be made.

Scattering from a polycrystalline sample takes place into Debye-Scherrer cones with the $k$ direction as axis and semiangles $2\theta$ defined by $\sin\theta = \tau/2k$. The total cross section associated with each cone is [from (3.15)]

$$\sigma(hkl) = \frac{4\pi^3}{k^2} \cdot \frac{N}{v_0} \cdot \frac{j_{(hkl)}}{\tau} |F_{(hkl)}|^2 \tag{3.21}$$

where $j_{(hkl)}$, the multiplicity, is the number of $\tau$ vectors with the magnitude $|\tau|$ and $|F_{(hkl)}|$ is the mean value of $F_N(\varkappa)$ for just these vectors. The counter observes only a fraction $l/2\pi r(\sin 2\theta)$ of the complete cone where $l$ is the height of the counter and $r$ the sample-counter distance. Thus,

$$I = I_0 \frac{\lambda^3 l}{8\pi r} \cdot V \frac{j_{(hkl)} N_c^2 |F_{(hkl)}|^2}{\sin\theta \sin 2\theta} \tag{3.22}$$

where $V$ is the volume of the sample and $N_c$ is the number of unit cells per unit volume.

## 3.3 Magnetic Scattering

The general cross section for the scattering of neutrons by magnetic interactions, derived from the master formula [Eq. (3.4)] and the interaction potential between a neutron and an electron with spin and momentum is derived by *Marshall* and

*Lovesey (1)*. We are interested in elastic scattering from spin-only systems where, for discussing magnetic scattering at low $\varkappa$ at least, residual orbital effects introduced via spin-orbit coupling may be accounted for by a factor $g/2$ associated with the form factor, where the $g$-factor may be determined by magnetic resonance measurements.

**Paramagnets.** In the absence of magnetic fields, there is no coherent scattering from paramagnets because of the random orientation of the magnetic ions. The differential cross section is

$$\frac{d\sigma}{d\Omega} = N\left(\frac{\gamma e^2}{mc^2}\right)^2 \frac{g^2}{4} \cdot \frac{2}{3} S(S+1) |f(\varkappa)|^2 \tag{3.23}$$

where $N$ is the number of magnetic ions of total spin $S$, m the mass of the electron and $c$ the velocity of light. $\gamma = -1 \cdot 91$ is the gyromagnetic ratio of the neutron[10]. $(\gamma e^2/mc^2)^2 = 0.29 \times 10^{-24}$ cm$^2$.

In a magnetic field at low temperatures a paramagnet will be significantly polarized in an applied field of several kgauss. The cross section is then

$$\frac{d\sigma}{d\Omega} = \left(\frac{\gamma e^2}{mc^2}\right)^2 \left[\frac{g}{2} f(\varkappa)\right]^2 \left\{\left(1 - \hat{H}_z^2\right) \langle S^z\rangle^2 N \frac{(2\pi)^3}{v_0} \sum_{\tau} \delta(\varkappa - \tau) \right.$$
$$+ N\left[\frac{1}{2} S(S+1) + \frac{1}{2}\langle (S^z)^2\rangle - \langle S^z\rangle^2 + \right. \tag{3.24}$$
$$\left.\left. + \hat{H}_z^2\left\{\frac{1}{2} S(S+1) - \frac{3}{2} \langle (S^z)^2\rangle + \langle S^z\rangle^2\right\}\right]\right\}$$

$\hat{H}_z = \hat{\boldsymbol{H}} \cdot \hat{\varkappa} = \cos\alpha$, where $\alpha$ is the angle between the magnetic field direction and the scattering vector. $\langle S^z\rangle$ and $\langle (S^z)^2\rangle$ are given by the Boltzman averages[11]. The first term describes the coherent scattering, important for a paramagnetic salt, which in a magnetic field becomes effectively ferromagnetic, [e.g., $K_2NaCrF_6$ where the $Cr^{3+}$ form factor was determined using polarized neutrons on a paramagnetic sample aligned in a magnetic field at 4.2 K (22)]. For such paramagnets and also for ferromagnets the magnetic unit cell is identical to the crystallographic unit cell. The coherent scattering is zero for zero applied field and also for paramagnetic ions randomly doped into a diamagnetic lattice (and for $\hat{H}_z^2 = 1$ with the magnetic field applied along the scattering vector).

---

[10] The magnetic moment of the neutron $\left(s = \pm \frac{1}{2}\right)$ is $2\gamma\mu_N \cdot \boldsymbol{s}$ where $\mu_N$ is the nuclear magneton $(e\hbar/2m_p c)$.

[11] $\langle S^z\rangle = \sum_{m=-S}^{+S} m \exp(mg\mu_B H/k_B T) \Big/ \sum_{m=-S}^{+S} \exp(mg\mu_B H/k_B T)$

$\langle (S^z)^2\rangle = \sum_{m=-S}^{+S} m^2 \exp(mg\mu_B H/k_B T) \Big/ \sum_{m=-S}^{+S} \exp(mg\mu_B H/k_B T).$

The second term describes the incoherent processes and is not zero for a random atomic distribution, giving rise to diffuse scattering[12]. Both coherent and incoherent terms can be separated from nuclear and other nonmagnetic terms by varying $\hat{H}_z^2$. For diffuse scattering measurements it is convenient to count with $\hat{H}_z^2 = 1$ (field along the scattering vector) and field off ($\langle \hat{H}_z^2 \rangle = 1/3$). In such experiments the cryostat-magnet arrangement rotates in a $\theta - 2\theta$ relation with the counter (Fig. 7). The difference cross section is

$$\frac{d\sigma}{d\Omega} = N\left(\frac{\gamma e^2}{mc^2}\right)^2 \frac{g^2}{4}\left\{\frac{S(S+1)}{3} - \langle (S^z)^2 \rangle\right\} |f(\varkappa)|^2 \tag{3.25}$$

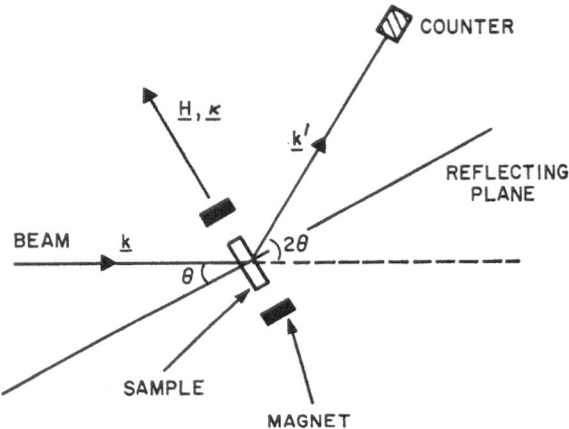

Fig. 7. Magnetic field-counter arrangement in a magnetic switching experiment [Eq. (3.25)]

**Magnetically Ordered Systems.** For magnetically ordered systems the discussion is analogous to that for nuclear scattering. By comparison with (3.15) the magnetic Bragg cross section is

$$\left(\frac{d\sigma}{d\Omega}\right)_{coh} = \frac{N(2\pi)^3}{v_0}\left(\frac{\gamma e^2}{mc^2}\right)^2 \sum_{\tau} \delta(\varkappa - \tau)$$
$$\times \left|\sum_j e^{i\varkappa \cdot r_j} \frac{g_j}{2} \langle S_j \rangle f_j(\varkappa) q_j e^{-W_j(\varkappa)}\right|^2 . \tag{3.26}$$

---

[12] We prefer to reserve the term incoherent scattering for the isotopic and nuclear spin processes already described. Magnetic diffuse scattering is not isotropic, having a form factor dependence. Nuclear diffuse scattering, arising from defects in the crystal lattice due to the effects of irradiation or from non-stoichiometry is also frequently angular dependent and provides information on short range atomic ordering (54).

The sum is over the magnetic ions in the magnetic unit cell. The $j$th ion has effective spin $\langle S_j \rangle$, form factor $f_j(\varkappa)$ and Debye-Waller factor $\exp[-W_j(\varkappa)]$[13]. $q_j$ is an orientation vector describing the direction of a unit vector $\hat{\eta}_j$ in the direction of the spin $\langle S_j \rangle$ relative to $\hat{\varkappa}$.

$$q_j = \hat{\varkappa}(\hat{\varkappa} \cdot \hat{\eta}_j) - \hat{\eta}_j \qquad (3.27)$$

$q_j$ is perpendicular to $\hat{\varkappa}$ and of magnitude $\sin\alpha$ where $\alpha$ is the angle between $\hat{\eta}_j$ and $\hat{\varkappa}$. Thus, only the atomic spin component perpendicular to $\varkappa$ is effective in scattering neutrons.

By analogy with nuclear scattering we may define a magnetic scattering length $p$ where

$$p_j = \left(\frac{\gamma e^2}{mc^2}\right) \frac{g_j}{2} \langle S_j \rangle f_j(\varkappa) \qquad (3.28)$$

giving a magnetic structure factor

$$F_M(\varkappa) = q_j p_j \sum_j e^{i\varkappa \cdot r_j} e^{-W_j(\varkappa)} . \qquad (3.29)$$

For collinear single spin-axis antiferromagnets, with only one type of magnetic ion, which are of principal interest in this discussion, the intensity will be proportional to $|F_M(\varkappa)|^2$ where

$$|F_M(\varkappa)|^2 = |q|^2 p^2 \left| \sum_j (\pm) e^{i\varkappa \cdot r_j} \right|^2 e^{-2W(\varkappa)} \qquad (3.30)$$

and

$$|q|^2 = 1 - (\hat{\varkappa} \cdot \hat{\eta})^2 \qquad (3.31)$$

The $(\pm)$ refers to the direction of the spin (up or down). In a powder experiment $|q^2|$ must be averaged over equivalent reflections (55).

At the Néel temperature (3.26) will become zero (for $T > T_N$, $\langle S_j \rangle = 0$) and paramagnetic scattering (3.23) is observed at higher temperatures[14]. The magnitude of $\langle S_j \rangle$ below $T_N$ should follow a Brillouin function and the Néel temperature may be determined from the temperature dependence of the intensity of a magnetic reflection.

The type of ordering or configuration of the spins [the array of plus and minus signs in (3.30)] is generally fairly readily determined in powder diffraction by observing which reflections are, in fact, observed. Often the magnetic unit cell of antiferromagnets is larger than the crystallographic cell. Except in certain

---

[13] It is assumed that the nuclear and electronic Debye-Waller factors are equal. *Moon et al.* (27) have confirmed this within the experimental error in their study on gadolinium.

[14] Short range magnetic order above $T_N$ may however result in residual intensity at the Bragg reflections.

cases, however, where the moments point along directions of high symmetry, $\langle |q^2| \rangle$ is often insensitive to the direction of $\boldsymbol{\eta_j}$, and single crystal experiments are generally needed to determine spin orientations. As already mentioned most elastic magnetic scattering experiments have been concerned with the details of spin configurations and orientations (the magnetic structure) and their interpretation, rather than with the accurate measurement of $\langle S_j \rangle$ and the detailed shape of $f(\boldsymbol{\varkappa})$ of interest here.

## 3.4 Polarized Beam Experiments

The form factor dependence for magnetic scattering entails a sharp decrease in magnetic scattering intensity with increasing scattering angle. This is most disadvantageous for scattering from powders where only a small section of each Debye-Scherrer cone is observed. Two extremes are shown in Fig. 8 and 9. For $BaTbO_3$ (28) ($S = 7/2$) the magnetic scattering cross section is very large and the low angle (111) and (311) reflections are more intense than any nuclear reflections (Fig. 8). In this case, and for MnO (48) ($S = 5/2$) it was possible to measure form factors with reasonable accuracy to $\sin\theta/\lambda \approx 0.5 Å^{-1}$ (where $f(\boldsymbol{\varkappa}) \approx 0.1$). This was because of the high $<S>$ and because the simple structures with minimum overlap between nuclear and magnetic reflections allowed an accurate estimation of the magnetic intensities. For more complicated structures or lower $<S>$ this has not been generally possible. For example for NiO (56) ($S = 1$) even the lowest angle magnetic reflections are weak compared to the nuclear intensities (Fig. 9) and only the first three or so may be accurately determined.

In a single crystal experiment weaker reflections may be measured more easily and the form factor for $Ni^{2+}$ in NiO was one of the first ionic form factors to be accurately determined (27).

Significant problems often remain however, both in measuring very weak magnetic reflections and in separating magnetic from nuclear intensities. More accurate

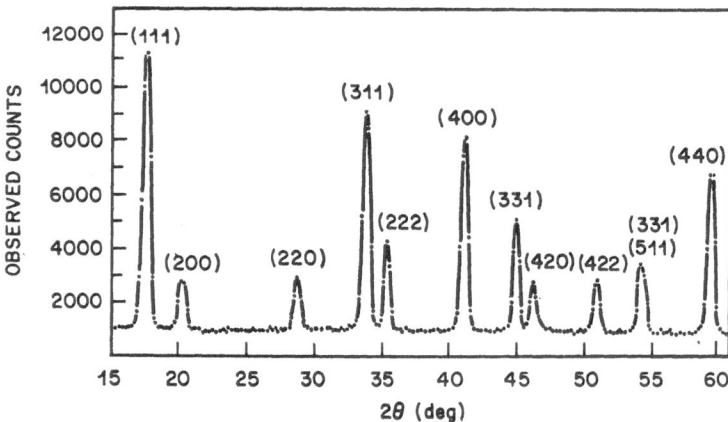

Fig. 8. Neutron diffraction pattern of $BaTbO_3$ at 4.2 K. The (111) reflection is entirely magnetic in origin and the other ($hkl$) all odd reflections have only a small nuclear contribution [Ref. (28)]

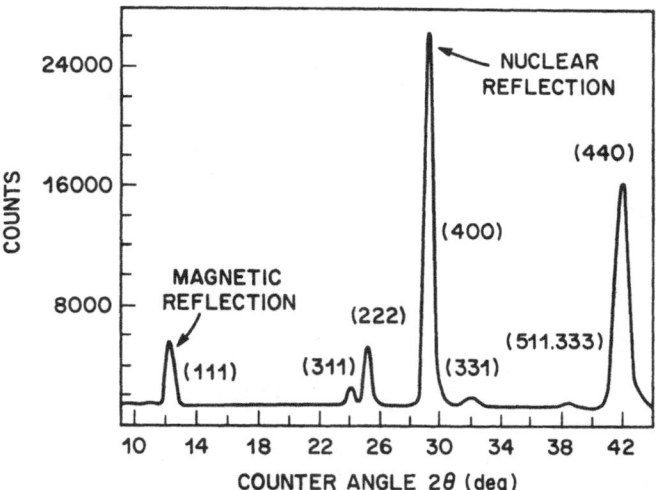

Fig. 9. Neutron diffraction pattern of NiO at 4.2 K. The magnetic reflections $[(hkl)$ all odd] are much weaker than the nuclear reflections [Ref. (29)]

estimation of magnetic scattering, especially at high $\varkappa$, where $f(\varkappa)$ is small, may sometimes be made using polarized neutrons. Accurate data at high angle are necessary if spin density distributions are to be determined by Fourier transform procedures.

When the spins are all parallel or anti-parallel to a given direction, the cross section is (1)

$$\frac{d\sigma}{d\Omega} = \left| \sum_{l} e^{i\varkappa \cdot l} \right|^2 |F_N(\varkappa)|^2 + N \sum_{r} (\overline{b_r^2} - \bar{b}_r^2)$$

$$+ \left( \frac{\gamma e^2}{mc^2} \right) \left| \sum_{l} e^{i\varkappa \cdot l} \right|^2 \left\{ \frac{g}{2} f(\varkappa) <S> P \cdot \hat{\varkappa} \times (\hat{\eta} \times \hat{\varkappa}) \times \sum_{r, r'} 2\bar{b}_r \cos [\varkappa \cdot$$

$$\cdot (r - r')] + \left( \frac{\gamma e^2}{mc^2} \right) \left| \frac{g}{2} f(\varkappa) \right|^2 <S>^2 |q|^2 \left| \sum_{r} e^{i\varkappa \cdot r} \right|^2 \right\} \qquad (3.32)$$

The first two terms are the nuclear coherent (3.15) and incoherent scattering (3.9) and the last term is the purely magnetic scattering (3.26). The third term is an interference term between nuclear and magnetic scattering and is zero if $\eta || \varkappa$, if the scattering is purely nuclear or purely magnetic, and if $P = 0$. $P$ describes the polarization of the incident beam $(0 \le |P| \le 1)$.

For ferromagnets, nuclear and magnetic reflections occur at the same $\tau$, and $\hat{\eta}$ may be established by an applied magnetic field. In particular, for $\hat{\eta}$ perpendicular to $\hat{\varkappa}$

$$P \cdot \hat{\varkappa} \times (\hat{\eta} \times \hat{\varkappa}) = P \cdot \hat{\eta} \qquad (3.33)$$

31

and $|q|^2 = 1$. For a Bravais lattice in this case

$$\frac{d\sigma}{d\Omega} = \left| \sum_l e^{i\varkappa \cdot l} \right|^2 \{\bar{b}^2 + 2\bar{b}pP \cdot \hat{\eta} + p^2\} \tag{3.33a}$$

where $p$ is the magnetic scattering length and nuclear incoherent scattering is neglected. For $P$ parallel or anti-parallel to $\hat{\eta}$ $(P \cdot \hat{\eta} = \pm 1)$

$$\frac{d\sigma^+}{d\Omega} = \left| \sum_l e^{i\varkappa \cdot l} \right|^2 (b + p)^2$$

and $\tag{3.34}$

$$\frac{d\sigma^-}{d\Omega} = \left| \sum_l e^{i\varkappa \cdot l} \right|^2 (b - p)^2.$$

The ratio of the Bragg intensities for positive and negative polarizations (the flipping ratio, $R$) is then

$$R = \frac{(1 + p/b)^2}{(1 - p/b)^2} = \frac{1 + \Gamma}{1 - \Gamma} \tag{3.35}$$

where $\Gamma$ is the ratio of the magnetic to nuclear scattering amplitudes.

The advantage of using polarized neutrons to determine weak magnetic reflections is clearly apparent. For example, if $\Gamma = 0.01$ then the magnetic contribution to the intensity in an unpolarized beam experiment is 0.01 percent, but $R \approx 1.04$, i.e., there is a 4 percent effect on changing the incident neutron polarization. It is necessary to know the nuclear scattering amplitude accurately if an accurate magnetic amplitude is to be obtained and extinction corrections in particular must be accurately performed.

A polarized beam apparatus is shown schematically in Fig. 10. The polarized beam is produced by reflection from a crystal for which $b = p$ $[d\sigma^-/d\Omega = 0,$ Eq. (3.34)]. The (111) plane of $Co_{0.92}Fe_{0.08}$ is frequently used, and polarization efficiencies of $\geq 99$ percent may be obtained. The polarization may be reversed

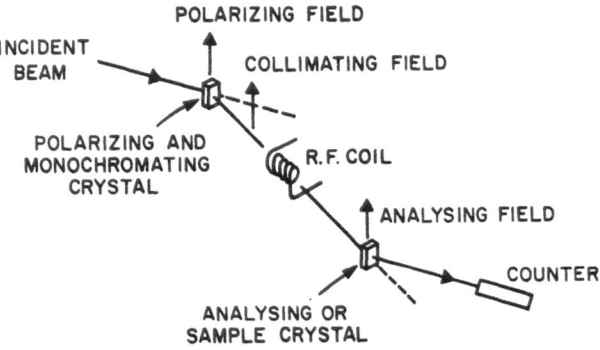

Fig. 10. Schematic diagram of a polarized beam apparatus

with an efficiency of $\geq 99$ percent by applying a radio-frequency field at right-angles to the neutron polarization and matched in frequency to the Larmor precessional frequency of the neutron spin in a uniform magnetic field. Using such an r.f. flipper is easier than reversing the field at the sample; unless the field reversal region is very small the neutron polarization will follow the field reversal. Corrections for imperfect polarization, imperfect reversal and depolarization by passage through the sample must be made and are described for example by *Moon et al. (21)*.

As mentioned above polarized beam methods have been most valuable in the determination of the spin distributions in ferromagnetic metals. Application to ionic materials has been limited because these are frequently antiferromagnetic with magnetic unit cells larger than the crystallographic unit cells so that magnetic and nuclear reflections often do not occur at the same $\tau$. Also the moments cannot be aligned by an external field. But paramagnets become effectively ferromagnetic in an applied field and $K_2NaCrF_6$ has recently been investigated by polarized neutrons *(22)*. It is likely that this type of experiment in particular, especially using polarized beam apparatus at high flux reactors, will provide in the next few years a great deal of interesting information about bonding and spin distributions for a wide variety of metal complexes. There are also some circumstances in which useful information may be obtained from antiferromagnets and experiments on $MnF_2$ *(57)* and $MnCO_3$ *(58)* are described below.

## 3.5 Polarization Analysis

The conventional polarized beam experiment is useful only for systems with polarization dependent cross sections. More information can often be gained if the polarization of the scattered neutrons is measured relative to the incident polarization (polarization analysis). This may be done in a triple-axis mode with a polarization sensitive analyzing crystal (Fig. 11). The technique was introduced experimentally by *Moon et al. (59)*. In their apparatus the magnetic field at the sample may be rotated about a horizontal axis. With the sample field vertical the neutron polarization remains vertical, but with the sample field horizontal the neutron polarization at the sample becomes horizontal (a reverse rotation occurs after the sample so that the polarization at the analyzing crystal is again vertical). Thus $P$ can be either parallel or perpendicular to the scattering vector $\varkappa$.

Four cross sections may be determined:

| Incident Polarization | Analyser | Cross Section Measured |
|---|---|---|
| + (1st flipper off) | + (2nd flipper off) | + + Non spin-flip Scattering |
| — (1st flipper on) | + (2nd flipper off) | —+ Spin-flip Scattering |
| + (1st flipper off) | — (2nd flipper on) | +— Spin-flip Scattering |
| — (1st flipper on) | — (2nd flipper on) | — — Non spin-flip Scattering |

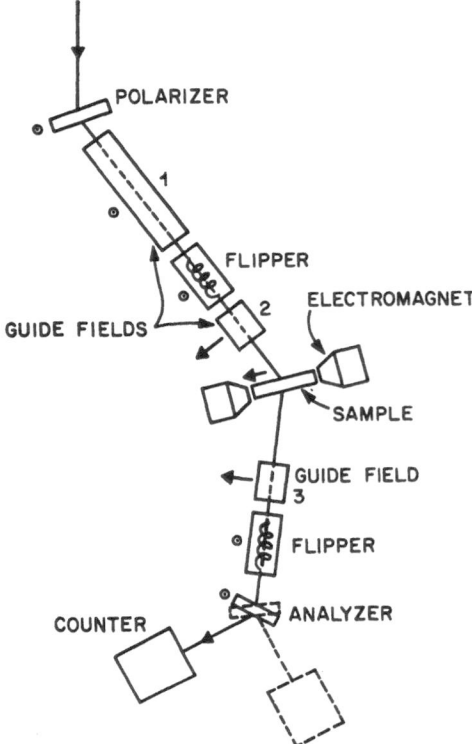

Fig. 11. Experimental layout for polarization analysis experiments [from Ref. (59)]. The magnetic field directions sensed by the neutrons are indicated

A comprehensive discussion of polarization analysis is given by *Marshall* and *Lovesey* (1). We shall only quote the expressions given in Ref. (59) which demonstrate the contributions to the various cross sections.

In the case of an incident polarized beam the master formula (3.4) may be written

$$\frac{d^2\sigma^{ss'}}{d\Omega dE} = \sum_\lambda p_\lambda \sum_{\lambda'} \frac{k'}{k} \left| < \lambda' \left| \sum_i e^{i\varkappa \cdot r_i} U_i^{ss'} \right| \lambda > \right|^2$$
$$\cdot \delta\left( \frac{\hbar^2}{2m_n} (k'^2 - k^2) + E_{\lambda'} - E_\lambda \right) \tag{3.36}$$

where the sum is over atomic sites and

$$U_i^{ss'} = \langle s' | (b_i - p_i \boldsymbol{S}_{\perp i} \cdot \boldsymbol{\sigma} + B_i \boldsymbol{I}_i \cdot \boldsymbol{\sigma} | s \rangle \tag{3.37}$$

is an atomic scattering amplitude describing a change in neutron spin state from $|s\rangle$ to $|s'\rangle$. $b_i$ and $p_i$ are the nuclear coherent and magnetic scattering amplitudes and $B_i$ is the spin-dependent nuclear amplitude[15]. $\boldsymbol{\sigma}$ is the neutron spin operator, $\boldsymbol{I}$ the nuclear spin operator, and

$$S_\perp = - q = \hat{\eta} - \hat{x}(\hat{x} \cdot \hat{\eta}) . \tag{3.38}$$

The four amplitudes $U_i^{ss'}$ are

$$
\begin{aligned}
U^{++} &= b - pS_{\perp z} + BI_z \\
U^{--} &= b + pS_{\perp z} - BI_z \\
U^{+-} &= - p(S_{\perp x} + iS_{\perp y}) + B(I_x + iI_y) \\
U^{-+} &= - p(S_{\perp x} - iS_{\perp y}) + B(I_x - iI_y)
\end{aligned}
\tag{3.39}
$$

where $z$ refers to the direction of neutron polarization.

From (3.39) we see that nuclear coherent scattering is always non spin-flip $(+ +, - -)$, as is nuclear incoherent scattering which is due to the random isotope distribution. This was demonstrated for polycrystalline nickel (59) (all nickel isotopes with significant abundance have $I = 0$). Magnetic and nuclear spin scattering is non spin-flip $(+ +, - -)$, if the effective spin components are along the neutron polarization direction, and spin flip $(+ -, - +)$ if the effective spin components are perpendicular to the polarization direction. Further, because only atomic spin components perpendicular to $\boldsymbol{x}$ are effective in scattering neutrons if follows that if the neutron polarization is along the scattering vector $(S_{\perp z} = 0)$ all magnetic scattering is spin-flip. This difference from nuclear scattering is the basis for distinguishing magnetic from nuclear scattering by polarization analysis. Elastic scattering only is considered here.

For nuclear incoherent scattering from a non-magnetic system with randomly oriented nuclear spins, the cross-sections per atom are independent of the neutron polarization direction and the spin-flip scattering cross section is twice that for non spin-flip scattering:

$$\frac{d\sigma^{++}}{d\Omega} = \frac{d\sigma^{--}}{d\Omega} = \frac{1}{3} B^2 I(I + 1)$$

and

$$\frac{d\sigma^{+-}}{d\Omega} = \frac{d\sigma^{-+}}{d\Omega} = \frac{2}{3} B^2 I(I + 1). \tag{3.40}$$

This was verified for vanadium (59).

The final polarization for paramagnetic scattering is

$$\boldsymbol{P'} = - \hat{x}(\hat{x} \cdot \boldsymbol{P}). \tag{3.41}$$

---

[15] $B_i = \dfrac{b_i^+ - b_i^-}{2I + 1}$. (Eq. 3.12).

Thus there is no scattered neutron polarization for an unpolarized incident beam. For a polarized beam, the scattered polarization is in the direction of the scattering vector. The partial cross sections per atom are [cf. (3.23)]

$$\frac{d\sigma^{++}}{d\Omega} = \frac{d\sigma^{--}}{d\Omega} = \frac{1}{3}\left(\frac{\gamma e^2}{mc^2}\right)^2 \frac{g^2}{4} S(S+1)f(\varkappa)[1 - (\hat{\varkappa} \cdot \boldsymbol{P})^2] \tag{3.42}$$

and

$$\frac{d\sigma^{+-}}{d\Omega} = \frac{d\sigma^{-+}}{d\Omega} = \frac{1}{3}\left(\frac{\gamma e^2}{mc^2}\right)^2 \frac{g^2}{4} S(S+1)f(\varkappa)[1 + (\hat{\varkappa} \cdot \boldsymbol{P})^2] . \tag{3.43}$$

For $\boldsymbol{P}=0$, the cross sections are equal, as is the case for $\boldsymbol{P}$ perpendicular to $\varkappa$ ($\hat{\varkappa} \cdot \boldsymbol{P}=0$). More importantly, for $\boldsymbol{P}$ parallel to $\varkappa$ ($\hat{\varkappa} \cdot \boldsymbol{P}=1$) the direction of polarization is reversed and the scattering is entirely spin-flip. Thus, paramagnetic

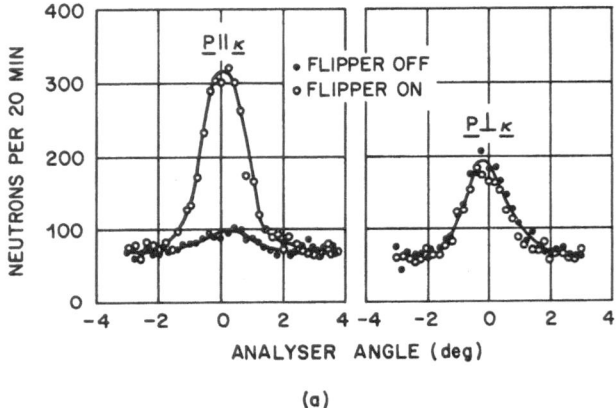

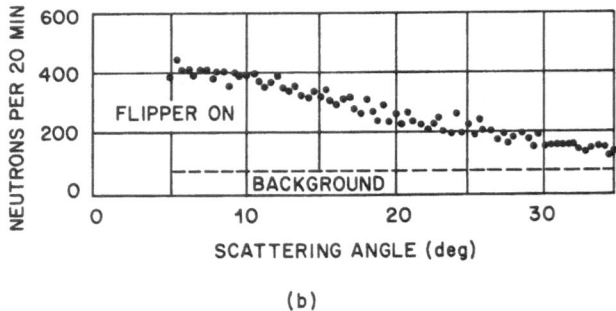

Fig. 12. Paramagnetic scattering from $MnF_2$ [from Ref. (59)]. The data were obtained by rocking the analyzer through the elastic position with fixed scattering angle. a) Comparison of scattering for $\boldsymbol{P}||\varkappa$ and $\boldsymbol{P}\perp\varkappa$. The "flipper-off" data are proportional to the $(++)$ cross section and the "flipper-on" data are proportional to the $(-+)$ cross section. b) Variation of paramagnetic scattering intensity with angle ($\boldsymbol{P}||\varkappa$)

scattering may be separated from all other diffuse and incoherent scattering processes except nuclear spin scattering. This is shown for $MnF_2$ in Fig. 12 (59). Nuclear spin scattering can also be separated by measuring spin-flip scattering for both $\hat{x} \cdot P = 1$ and $\hat{x} \cdot P = 0$. The two cross sections are respectively

$$\frac{d\sigma^{+-}}{d\Omega} = \frac{d\sigma^{-+}}{d\Omega} = \left(\frac{d\sigma}{d\Omega}\right)_{\text{para}} + \frac{2}{3}\left(\frac{d\sigma}{d\Omega}\right)_{\text{NS}}$$

$$\frac{d\sigma^{+-}}{d\Omega} = \frac{d\sigma^{-+}}{d\Omega} = \frac{1}{2}\left(\frac{d\sigma}{d\Omega}\right)_{\text{para}} + \frac{2}{3}\left(\frac{d\sigma}{d\Omega}\right)_{\text{NS}}$$

(3.44)

The study of paramagnetic scattering by polarization analysis will, hopefully, provide much interesting information on spin distributions and bonding. One study has already been made, on $Gd_2O_3$ (27). The normal corrections for beam polarization, flipper efficiency, multiple scattering and so on must be made. It is also important to check for residual short-range magnetic order.

Coherent magnetic scattering from antiferromagnets may also be separated from coherent nuclear scattering if $P$ is parallel to $\varkappa$, for in this case also nuclear scattering is non spin-flip and the magnetic scattering is spin-flip. Only a small field ($\approx 100G$) is necessary to guide the polarization at the sample. The non spin-flip nuclear scattering is given by Eq. (3.8) and the spin-flip magnetic cross sections are

$$\frac{d\sigma^{\pm\mp}}{d\Omega} = \sum_{ij} e^{i\varkappa \cdot (r_i - r_j)} p_i p_j^* [S_{\perp i} \cdot S_{\perp j} \mp i\hat{Z} \cdot (S_{\perp i} \times S_{\perp j}^*)].$$

(3.45)

This is the cross section already derived in the case of collinear single spin-axis systems. $\hat{Z}$ is a unit vector along the polarization direction.

This is particularly useful where magnetic and nuclear reflections occur at the same value of $\tau$, and especially for powder samples. The separation is shown for $\alpha$-$Fe_2O_3$ (Fig. 13). The technique has been used to establish the absence of magnetic order in $Ti_2O_3$, (60) and the presence of magnetic order, and the effective moment in $V_2O_3$, (14) and has potentially wide applicability to the determination of moment reductions and form factors in antiferromagnetic salts where nuclear and magnetic reflections often overlap.

A serious problem, however, for both paramagnetic and coherent magnetic scattering measurments, at least until higher reflectivity polarizers are available or alternative techniques of polarization analysis are developed, is the low intensity obtained after reflection from the sample as well as from two polarizing crystals. Even for location of apparatus at a high flux reactor, with current techniques the measurement of covalent spin reductions in powder samples can be made as efficiently by profile analysis of a conventional powder diffraction pattern as by polarization analysis. Polarized beam methods including polarization analysis will however, be essential for the determination of form factors.

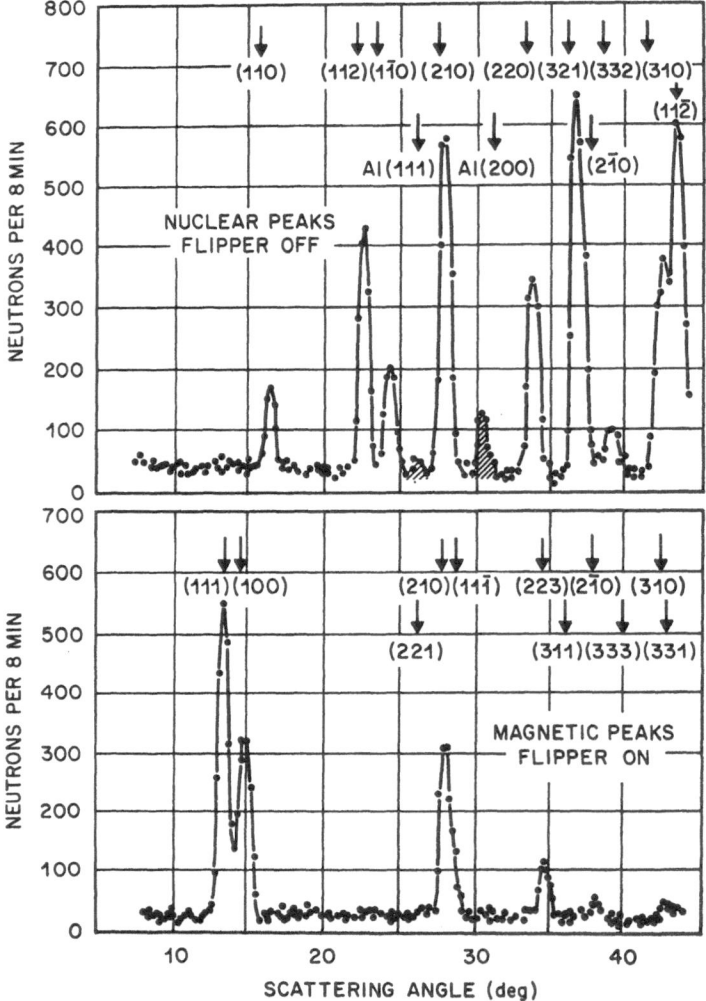

Fig. 13. $\alpha$-$Fe_2O_3$ powder pattern — separation of nuclear and magnetic peaks by polarization analysis $(P\|\varkappa)$; [from Ref. (59)]

## 3.6 Powder Diffraction Techniques

A typical powder diffraction apparatus is shown in Fig. 14. The desirable requirements of high beam intensity and high resolution are rather incompatible and conventional instruments often have fairly poor resolution. This is advantageous in experiments where peaks are well separated as count rates can be high, and was the situation for example, in the measurements on $Mn^{2+}$ and $Ni^{2+}$ rock-salt compounds (56, 61) and $Cr^{3+}$, $Fe^{3+}$ and $Mn^{4+}$ oxide perovskites (62), where only a few low-angle peaks were measured. High resolution is essential in profile analysis refinement, however, or when many intensities must be measured to

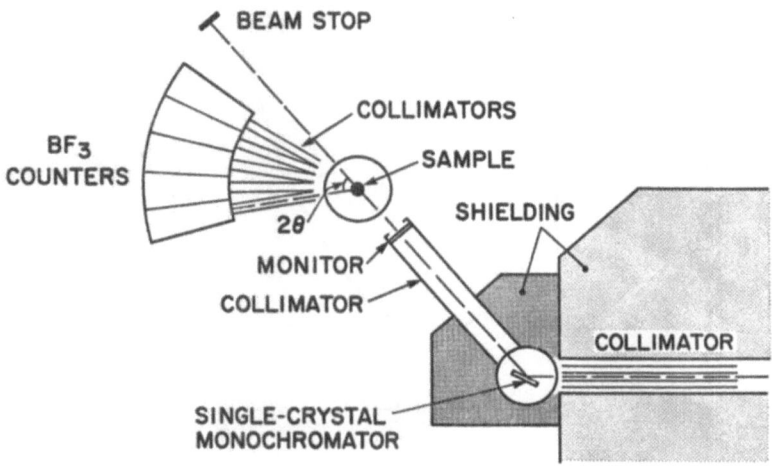

Fig. 14. Schematic diagram of a powder diffraction apparatus. A counter bank with five counters of fixed angular separation is shown. Use of a multicounter system allows more rapid data collection than with a single counter

perform a structure refinement. *Caglioti* (*63*) has discussed the collimation problems involved. The profile refinement procedure is very sensitive to errors in the peak positions and the scattering angle must be known to at least 0.1°. The PANDA diffractometer at Harwell on which several of the experiments described below have been performed uses a Moiré fringe method for positioning the counter. Generally the counter is programmed to make steps of 0.02° and to print out every 0.10° $2\theta$.

When accurate data are to be obtained to $> 100°$ $2\theta$ care must be taken in the choice of the Bragg angle of the monochromator. The typical variation of half width of a Bragg reflection with scattering angle is shown in Fig. 15 with the minimum occurring at the Bragg angle of the monochromator. The increase in half width is particularly severe at higher scattering angles and a high take-off angle (say 90°) is essential.

Because of the thermal distribution of neutron energies at the monochromator the reflection of lower wavelength ($\lambda/2$, $\lambda/3$, etc.) neutrons by higher order planes of the monochromator can be a problem. For the wavelengths often used ($1.0 - 1.5$Å) only $\lambda/2$ neutrons need be considered, but for longer wavelengths higher orders must also be taken into account. For most monochromator crystals used (Pb, Cu, etc.) $\lambda/2$ neutrons will be present and their effect must be subtracted from the measured intensities. Thus in the determination of the $Mn^{2+}$ form factor in MnO (*48*), where a copper monochromator was used, $\lambda/2$ peaks were present under all the magnetic reflections and became more significant as the magnetic reflections became weaker at higher angle. For the (55$\bar{3}$, 73$\bar{1}$) the $\lambda/2$ contribution was as large as 5%. The proportion of $\lambda/2$ neutrons in the beam was determined by measurement of the (110) intensity of yttria-stabilized cubic $ZrO_2$. This reflection is forbidden for primary neutrons but was

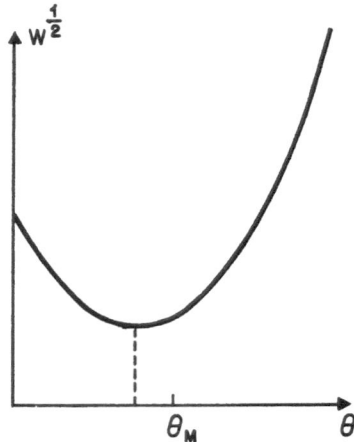

Fig. 15. Typical peak width at half-height $(W^{1/2})$ as a function of the Bragg angle $\theta$. The minimum occurs for $\theta$ close to the Bragg angle of the monochromator $\theta_M$

weakly observed by $\lambda/2$ scattering from the strong (220) reflection. $\lambda/2$ neutrons are not produced by reflection from odd index planes of crystals with the diamond structure [*e.g.*, for reflection from (111), the (222) reflection is forbidden] and although the reflectivity is not as high as for some metal monochromator crystals, a germanium monochromator may be used when data of high accuracy without $\lambda/2$ contamination are required. By the same token it is most satisfactory to use a vanadium sample container and a vanadium-tailed cryostat, instead of the more conventional aluminum, to avoid spurious reflections in the diffraction pattern [Eq. (3.14) and Section 3.1].

Because the observed covalent spin reductions are of the order of only 5—15%, with an intensity reduction of 10—30%, it is clear that data of high quality are required if the reduction is to be measured with any accuracy, particularly if the spin is relatively low as for $Cr^{3+}(S=3/2)$ and $Ni^{2+}(S=1)$ so that the magnetic intensities are low in the first place. For this reason it is important to eliminate as many other potential uncertainties as possible. Measurement is preferably carried out at 4.2K or lower to achieve magnetic saturation and a simple magnetic structure is desirable to avoid uncertainties in $|q|$.

As mentioned above, for experiments on several compounds with simple magnetic and nuclear structures the spin reduction was calibrated absolutely against a known nuclear intensity, avoiding the necessity to determine the scale factor. Variations in Debye-Waller factor were also negligible for fairly closely spaced peaks at low angle. A nuclear intensity of the same compound was used where the nuclear structure was particularly simple [*e.g.*, for NiO (Fig. 9) the magnetic (111) was calibrated against the nuclear (400) (*56*)], or if the nuclear structure was not exactly known a weighed amount of substance of known structure and scattering length could be added [*e.g.*, for $YFeO_3$ (Fig. 16) and other oxide perovskites the magnetic (011 + 101) was calibrated against the germanium (111) re-

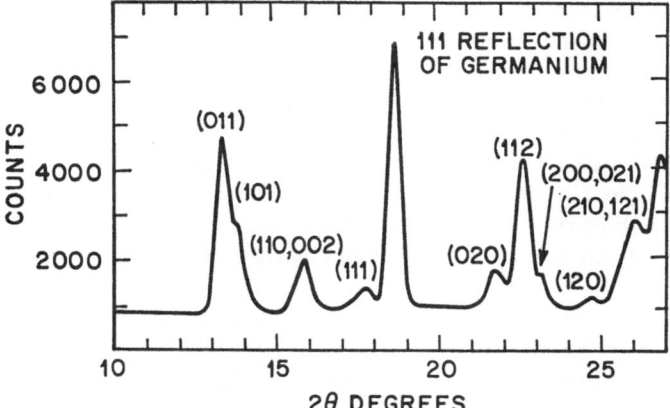

Fig. 16. Neutron diffraction pattern of YFeO₃. The lowest angle (011)+(101) peaks are predominantly magnetic in origin and were calibrated against the Ge (111) reflection. This only overlapped slightly with the small (111) nuclear reflection of YFeO₃ and the background could be accurately measured

flection (62)[16]]. These compounds were chosen so that overlapping peaks were not a problem and the background could be accurately determined. Standard deviations ($\sigma$) of peaks were always less than 1%. $\sigma$ is given by

$$\sigma = \sqrt{P + \left(\frac{n_p}{n_b}\sqrt{B}\right)^2} \qquad (3.46)$$

where $P$ is the total peak plus background, $B$ the total background under the peak, $n_p$ the number of points counted under the peak and $n_b$ the number of points taken to average the background. Several determinations of peak ratios were always made and it was generally found that variations due to sample alignment errors were greater than the statistical errors on each determination. Consequently an average was taken over several runs and the standard deviation of this value taken as the experimental error.

Such measurements give an accurate value of $\langle S \rangle$ although no indication of the shape of $f(\varkappa)$. It is thought to be a fairly good assumption that for the low angle magnetic peaks measured $f(\varkappa)$ is not greatly different from the free ion values [Fig. 5 and Ref. (64)]. This was verified for NiO (56) where the actual form factor was known (27). Application of the technique was primarily limited by the relatively small number of magnetically ordered 'ionic' materials with appropriately simple magnetic and nuclear structures.

The principal uncertainty with these, and all other determinations on anti-ferromagnets lies in the estimation of the zero-point spin deviation. Even at zero

[16]) The scattering length of the calibrant must of course be accurately known, as must those of the sample if an internal calibration is being made. This was well discussed by *Hutchings* and *Guggenheim* (82) in estimating the errors associated with the covalency parameter for Ni²⁺—F⁻ determined in KNiF₃.

degrees, zero point motion causes a reduction in $\langle S \rangle$ below the free ion value and this must be taken into account when determining the covalent reduction. Although quite accurate estimates have been made for a few compounds [e.g., for NiO and MnO (65),] estimates using the spin-wave theory of *Anderson* (66) or the perturbation theory of *Davis* (67) have generally to be used. At the present time, the lack of accurate calculations of zero point spin reductions is the greatest source of uncertainty in the covalency parameters determined for antiferromagnets. Nor is it known how the effect will vary with deviations from regular coordination. On the other hand there is no problem with ferromagnetic or paramagnetic materials.

Several antiferromagnetic materials studied are also weakly ferromagnetic due to spin canting (68) (Fig. 17). The canting angles which are generally observed, however, indicate that this effect causes an insignificant reduction in the sublattice magnetization. Nevertheless, by use of the polarized beam, *Brown* and *Forsyth* (58) were able to determine the $Mn^{2+}$ form factor in $MnCO_3$ by measurement of the ferromagnetic intensities arising from spin canting.

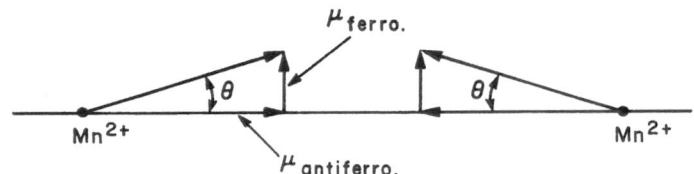

Fig. 17. Schematic diagram of relative spin orientations in a canted antiferromagnet such as $MnCO_3$

In two cases of materials with high spin and relatively simple nuclear and magnetic structures it was possible to measure the shape of $f(\varkappa)$ as well as $\langle S \rangle$. For $BaTbO_3$ (28) ($S = 7/2$) and MnO (48) ($S = 5/2$) an accurate separation of nuclear and magnetic intensities could be made, and a nuclear structure determination carried out at 4.2 K. From the scale factor determined, $\langle S \rangle f(\varkappa)$ could be determined for the magnetic reflections up to $\varkappa \approx 6.5 \text{Å}^{-1}$. The shape of $f(\varkappa)$ was determined by assuming the free ion value at the lowest angle reflection to find $\langle S \rangle$. The value of $\langle S \rangle$ determined for MnO (48) in this way agreed very well with the earlier value determined by calibration with one nuclear intensity (56). In these experiments the nuclear structure was determined by a least squares analysis based on intensities, the function minimized being $\sum \omega_i (I_{obs} - I_{calc})^2$ where $\omega_i = 1/\sigma_i^2$.

Such a refinement program was very useful in these cases but is in general of limited application, for it is only with very simple structures (both nuclear and magnetic) that a sufficient number of nuclear intensities can be accurately resolved at 4.2 K to provide a basis for refinement and the determination of the scale factor. A more general refinement procedure has been recently introduced (69) which fits the measured profile of the powder diffraction pattern rather than individual intensities or structure factors. With data of high resolution obtained over a wide angular range nuclear structure refinements may be performed with good precision even if thirty or more quantities are varied, and the results obtained appear to be of comparable precision to those determined by much lengthier single crystal methods [for a comparison see *e.g.*, Ref. *(70)*]. This method of refinement is particularly suited to neutron diffraction because the profile of a single powder diffraction peak is almost exactly Gaussian *(69)* in contrast to the more complicated line shapes observed in X-ray diffraction. Moreover magnetic and nuclear refinements may be performed simultaneously so that $\langle S_z \rangle$ can be obtained in situations where overlapping peaks do not allow a simple separation of nuclear and magnetic intensity. Application has recently been made *(71)* to octahedral and tetrahedral $Fe^{3+}$ in $Sr_2Fe_2O_5$ and to tetrahedral $Co^{2+}$ in $Co_3O_4$ and $CoRh_2O_4$ *(72)*. In most cases it will probably be difficult to determine form factor deviations from free ion behavior with any confidence because at higher angles where significant deviations of $f(\varkappa)$ may occur, the nuclear scattering will be dominant due to the drop in magnitude of $f(\varkappa)$. Nevertheless, this method allows a great number of antiferromagnetic compounds to be studied for which spin reductions could not otherwise be determined with any accuracy. As mentioned above, polarization analysis methods will be necessary for the accurate determination of form factors, but the nuclear structure must still be determined accurately to give the scale factor. For the routine measurement of spin reductions, which can be determined at the same time as the nuclear refinement, profile analysis of a conventional powder diffraction pattern, which does not suffer from the severe intensity losses of the polarization analysis techniques described above, seems to offer a considerable saving in time, especially when multiple-counter detection is employed. Accurate powder diffraction patterns may also be obtained using lower-flux reactors where it is not feasible to install a polarization analysis facility.

## 3.7 Spin Density Distributions from Single Crystal Data

Experiments with powders yield $\langle S_z \rangle$ and in certain unusual situations, significant data on the shape of $f(\varkappa)$ have been obtained. Even so, only a spherical average is generally determined which, although adequate to discuss the high spin $d^5$ and $f^7$ materials studied [MnO *(48)* and $BaTbO_3$ *(28)*], can conceal much interesting information.

Single crystal experiments are essential in the detailed investigation of spin density distributions. The anisotropy in the form factor at a given $\varkappa$ can be measured and if a fairly extensive set of information is collected the spin density may be determined by Fourier transformation techniques.

In such work, particularly in the case of metals, the spin density at any position in the unit cell cannot be uniquely assigned to a particular atom[17]. The discussion of magnetic scattering given above implicitly assumes nonoverlapping spins, and

---

[17] It is, however, possible to determine in some instances than spin density has been transferred to formally nonmagnetic atoms by observing a magnetic component for reflections which otherwise should arise from nuclear scattering only [*e.g.*, on $F^-$ in $MnF_2$ *(57)* and $O^{2-}$ in $Y_3Fe_5O_{12}$ *(73)* (see below)].

this approximation is probably valid in many cases of interest here. As discussed by *Marshall* and *Lovesey* (*1*) however, it is convenient to consider a more general case valid for both metals and ionic systems. If the spin density at position $r$ in the sample has magnitude $s(r)$, the magnetic cross-section is

$$\frac{d\sigma}{d\Omega} = \left(\frac{\gamma e^2}{mc^2}\right)^2 \left| \int d\boldsymbol{r}\ e^{i\varkappa \cdot \boldsymbol{r}} (\hat{\varkappa} \times s(\boldsymbol{r}) \times \hat{\varkappa}) \right|^2 \tag{3.47}$$

Because $s(r)$ is periodic

$$\int_V d\boldsymbol{r}\ e^{i\varkappa \cdot \boldsymbol{r}}\ s(\boldsymbol{r}) = \frac{2\pi^3}{v_0} \sum_{\tau} \delta(\varkappa - \tau) F(\tau) \tag{3.48}$$

where $F(\tau)$ is the magnetic unit cell vector structure factor. Inverting (3.48)

$$s(r) = \frac{1}{v_0} \sum_{\tau} e^{-i\tau \cdot \boldsymbol{r}}\ F(\tau) \tag{3.49}$$

$V$ is the sample volume and $v_0$ the unit cell volume. Conversely

$$F(\tau) = \int_{v_0} d\boldsymbol{r}\ e^{i\tau \cdot \boldsymbol{r}}\ s(\boldsymbol{r}) \tag{3.50}$$

The spin density in the unit cell is determined from (3.49)[18]. The majority of such experiments have been concerned with ferromagnetic metals and compounds. For these, and for paramagnetic materials the polarized beam technique gives $F(\tau)$ for $\tau \neq 0$ and the form factor can be normalized by a determination of the bulk magnetization [giving $F(0)$].

In addition to this measurement, great care must be taken in the diffraction measurements; in particular the nuclear structure factors must be accurately known to extract $F(\tau)$ accurately from the flipping ratio.

The use of a bulk magnetization measurement to normalize the form factor may be criticized (*74*), particularly in metallic systems where it may include the effects of conduction spin polarization which will not follow the form factor dependence obeyed by the localized spin density. An alternative procedure proposed by *Moon* (*74*) and others is to split the total magnetization into two parts

$$s(r) = s_0 + s'(r) \tag{3.51}$$

---

[18] We have considered only a spin component. In a general case there is also an orbital contribution to the observed structure factor so that the moment density will actually be determined. The orbital contribution is important for non-orbitally singlet transition metal ions and for rare earth metals and ions. The separation of orbital and spin components has been discussed by *Moon* (*74*) and *Marshall* and *Lovesey* (*1*).

where $s_0$ is the average magnetization and $s'(\mathbf{r})$ is a periodic function given by Eq. (3.49) but where the set of $F(\boldsymbol{\tau})$ does not include $F(0)$. The atomic moment $\mu_{\mathrm{a}}$ may then be obtained by integrating over an appropriate volume $V_{\mathrm{a}}$:

$$\mu_{\mathrm{a}} = \int s'(\mathbf{r}) \, d\tau - V_{\mathrm{a}} s'(\mathbf{r}_{\mathrm{B}}) \tag{3.52}$$

where $r_{\mathrm{B}}$ is located in the region of assumed constant magnetization density. *Moon* has shown (74) that this procedure gives atomic moments for Fe, hexagonal Co and Ni in good agreement with the $3d$-like moments obtained originally by fitting free-atom form factors to the data[19]). Also, a local moment integration for Gd (21) gave 6.44 $\pm 0.16$ $\mu_{\mathrm{B}}$, in excellent agreement with the expected value of 6.42 $\mu_{\mathrm{B}}$ for the $4f$ electrons at the temperature of the measurement. The total magnetic moment is 6.92 $\mu_{\mathrm{B}}$. The integration method has the advantage that no assumptions are made about the shape of the form factor.

In ionic crystals it seems reasonable that the background level should fall to zero at suitably chosen positions in the unit cell so that a moment integration may yield the local moment directly. Such a procedure was followed for integration around the octahedral and tetrahedral $Fe^{3+}$ sites in yttrium iron garnet (72), where the position of the yttrium ion was chosen as the backgound level. This technique is particularly useful in such cases where there is more than one type of magnetic ion per unit cell.

Because of the decrease of magnetic structure factors with increasing angle and the general use of $\sim 1$Å neutrons, magnetic intensities are not generally measured to $\sin\theta/\lambda > 1.0 - 1.2$. A data set thus determined will give rise to series termination errors in the transformation procedure and in particular $s(\mathbf{r})$ obtained from Eq. (3.49) will oscillate for large values of $r$. However, for large $\mathbf{r}$ the spin density will not be changing rapidly and this difficulty can be overcome by asking for lower resolution and computing the average value of $s(\mathbf{r})$ over a small volume. For a cube with edges parallel to the cell axes and of length $2\delta$ the spin density average over this volume is $\overline{s(\mathbf{r})}$ where

$$\overline{s(\mathbf{r})} = \frac{1}{v_0} \left(\frac{a}{2\pi\delta}\right)^3 \sum_{t_1 t_2 t_3} \left(\frac{F(\boldsymbol{\tau})}{t_1 t_2 t_3}\right) \sin\left(\frac{2\pi t_1 \delta}{a}\right) \sin\left(\frac{2\pi t_2 \delta}{a}\right) \sin\left(\frac{2\pi t_3 \delta}{a}\right) \tag{3.53}$$

where $a$ is the lattice parameter and $t_1$, $t_2$ and $t_3$ are reciprocal lattice indices. $\overline{s(\mathbf{r})}$ should be almost independent of the choice of $\delta$ for a given $\mathbf{r}$. The factor $(t_1 t_2 t_3)^{-1}$ makes this series converge much more rapidly than that for $s(\mathbf{r})$. Such an averaging procedure is generally performed. Averaging methods are discussed by *Moon* (74) and *Marshall* and *Lovesey* (1).

Single crystal data have provided much new and interesting information for metallic systems in particular, and for some ionic systems, where, for example, form factor deviations from free ion calculations for $Ni^{2+}$ (27) and tetrahedral $Fe^{3+}$ (73) have been revealed. Also, the transformed spin density [Eq. (3.49)] is

---

[19]) The bulk magnetization moments are lower due to the presence of negative conduction spin density.

particularly interesting in demonstrating the directional properties of magnetic orbitals and in showing spin density transferred to anion groups. Thus in NiO (27) the $e_g$ nature of the spin density is apparent and in $K_2NaCrF_6$ (22) the $t_{2g}$ shape. In this compound spin transferred to $F^-$ is seen, as is spin density on the $CO_3^{2-}$ ions in $MnCO_3$ (58). It is not clear, however, how deeply one may probe into the effects of bonding using Fourier methods, where series-termination errors can give rise to effects as large as those of covalency, particularly for higher values of $\sin\theta/\lambda$. This problem was encountered in the work on $K_2NaCrF_6$ (22) where spherically separated form factors were determined from the data by a Fourier procedure for comparison with form factors calculated from free ion wavefunctions for $Cr^{3+}$ and $F^-$. The more lengthy procedure of calculating the magnetic scattering amplitudes directly from the model would avoid the effects of series termination. Such problems are likely to be more severe where orbital moment density is also present, as in the study of ions such as $V^{3+}$, $Fe^{2+}$ and $Co^{2+}$ in octahedral coordination (75).

### 3.8 Diffuse Scattering Apparatus

An investigation of a ruby single crystal (20) was made at low $\varkappa$ to measure the ligand forward peak [Eqs. (2.26), (2.27)]. The apparatus used was qualitatively different from the diffractometers discussed above and is briefly described. Wavelengths between 4Å and 7Å are selected by a simple time-of-flight procedure. A roughly collimated beam is passed through polycrystalline beryllium to filter out neutrons of wavelength <3.95Å (Fig. 18). The filtered beam is pulsed by a five-hole rotor which provides a time-base for the wavelength selection. The resolution is determined by the length of the counting time. This procedure provides a crude discrimination against inelastically scattered neutrons. In this apparatus a bank of thirteen fixed counters measured neutrons scattered in a vertical plane. Two samples could be compared by utilizing a sample changer in which two samples could be moved alternately into the beam along a horizontal axis. Cylindrical samples ($\sim$1 cm diameter $\times$ 2 cm long) were mounted in aluminum cans. The intensity was put on to an absolute basis by calibration with a polycrystalline vanadium cylinder of similar dimensions.

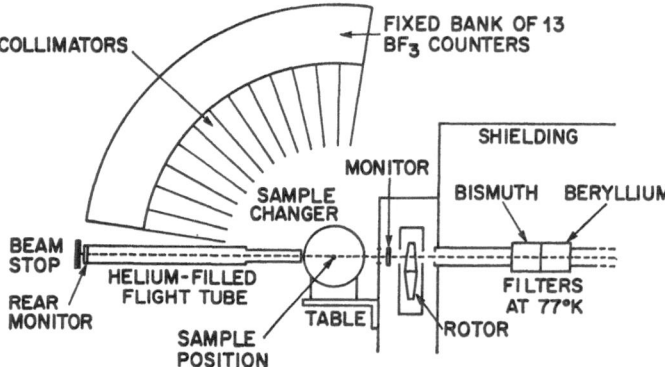

Fig. 18. Schematic diagram of the long-wavelength elastic diffuse scattering apparatus used to study the paramagnetic scattering from ruby [Ref. (20)]

# 4. Results

The data obtained so far are surveyed and comparisons made where appropriate with results from other techniques, especially resonance methods, and with calculations. Most results are for the spin-only $d^3$, $d^5$ and $d^8$ $3d$ transition metal ions and these are discussed first. It is convenient to choose the order $d^8$, $d^3$ and $d^5$. A few results obtained on other $3d$ ions are also mentioned, where appropriate, under these headings. Data on rare earth ions are then discussed.

## 4.1 $d^8$ Ions — $Ni^{2+}$

The $d^8$ octahedral system is apparently one of the simplest transition metal systems to analyse by the MO theory described above. Resonance data give $f_\sigma$ and $f_s$ and diffraction data $f_\sigma + f_s$. Both techniques seemed to give similar data for the oxides and fluorides studied without the apparent complications of spin polarization observed in $d^3$ complexes, or the 'anomalous' bonding parameters and form factors observed for $Mn^{2+}$ ($d^5$), and the form factor expansion observed in NiO (27) was in fairly good agreement with the simple theory (Section 2.2) and with more detailed calculation (26, 76). The recent ENDOR result (77) for $f_\sigma$ in $^{17}O$-doped MgO is much greater than that observed (27, 56) for NiO by neutron diffraction however. This confuses a situation that has long been thought to be fairly straightforward[20]).

First, however, let us discuss the classic study of NiO (27) by single crystal diffraction methods. A direct Fourier projection of the data showed that the unpaired spin density had $e_g$ symmetry (Fig. 19), a gratifying experimental demonstration of the correctness of some of the assumptions of the ligand field theory. NiO (and MnO) has magnetic ordering of the second kind (Fig. 20), the result of the dominant 180° superexchange interactions, with a slight rhombohedral distortion below $T_c$. The magnetic unit cell has a side twice that of the chemical cell and the magnetic reflections occur for $hkl$ all odd (cubic indexing). The spins lie in the (111) planes. The data were collected at room temperature ($T_N = 257$ °C) and the magnetic saturation was found to be 0.95 $\pm 0.05$ by measuring the intensity of the (111) reflection at 4.2K and room temperature. A moment of $gS = 1.81 \pm 0.20$ $\mu_B$, considerably lower than expected, was obtained and the form factor was found to be expanded 17% relative to the calculated (79) free ion value in the $\sin\theta/\lambda$ direction. These data provided the stimulus for the molecular orbital theory (26) of covalent effects in neutron diffraction. The spin reduction in polycrystalline NiO was remeasured by *Fender et al.* (56) and

---

[20] "It would appear, therefore, that NiO is fairly straightforward and that the results are understood," *Rimmer* (1968) (78).

47

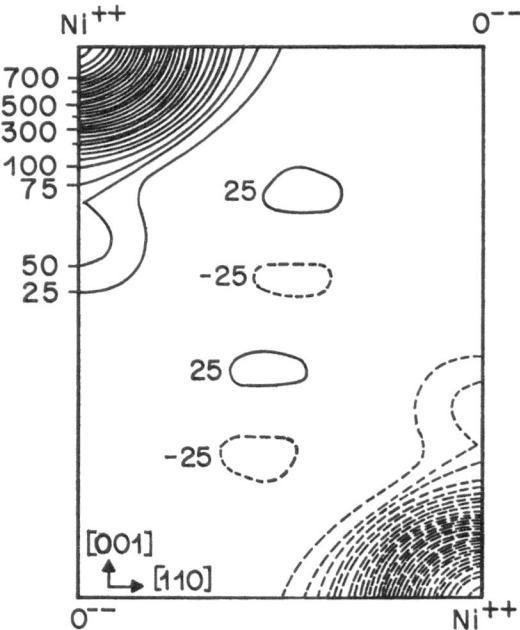

Fig. 19. The relative unpaired spin density in antiferromagnetic NiO. The solid and dashed contours denote positive and negative density respectively. The circle-like contours in the center arise from series termination errors [after Ref. (27)]

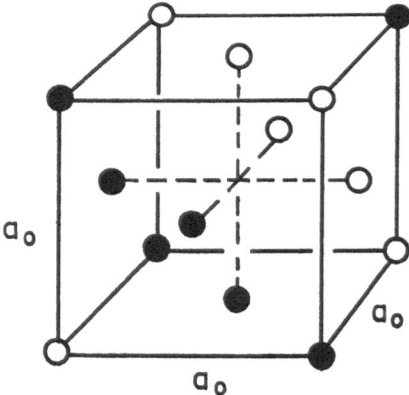

Fig. 20. Face-centered-cubic antiferromagnetic ordering of the second kind as exhibited by MnO and NiO. Only the metal atoms are shown and the black and open circles indicate spins of opposite orientation. $a_0$ denotes the crystallographic (cubic) cell above $T_N$

$f_\sigma + f_s$ found to be 3.8 ±0.2% (using $g = 2.23$) in good agreement with the value of 4.1% derived from Alperin's data. In this instance, knowing the correct form factor a test of the approximation that the free ion form factor may be used at low angles was possible. If the free ion, rather than Alperin's form factor was taken, $(f_\sigma + f_s)$ was found to be 3.1 $\pm 0.2\%$ which is reasonably close to the value quoted.

The $Ni^{2+}$ form factor was first calculated by *Hubbard* and *Marshall* (26) who included the small effects due to the small orbital moment (80) and the spin polarization of the occupied inner shells (81). Quite good agreement was obtained with the experimental curve (Fig. 21) after scaling to allow for the covalent spin reduction. A more recent calculation by *Soules* and *Richardson* (76) using unrestricted Hartree-Fock wave functions for the $NiF_6^{4-}$ cluster gave essentially the same result, with the agreement being best at lower scattering angles (orbital moment and spin polarization corrections were not made in this case). Unfortunately there have been no form factor determinations for $Ni^{2+}$ in any environment since the first experiment of *Alperin* (27).

Moment reductions have been measured for NiO (56) and for $KNiF_3$ (82) by powder diffraction[21] and ligand hyperfine interactions in $KNiF_3$ (83) and $KMgF_3$ (84) by NMR and ESR, and recently in $^{17}O$-doped MgO by ENDOR (77). The results are given in Table 3. As already mentioned, until the determination of $f_\sigma = 8.5\%$ in MgO, it seemed clear from both the neutron and the resonance data that for this divalent ion, oxides showed similar covalency to fluorides, with a ligand-to-metal transfer of approximately $0.2e$ [Eq. (2.23)] ($Mn^{2+}$, but not $Fe^{3+}$ data could be similarly interpreted).

Table 3. Spin transfer coefficients for $Ni^{2+}$

| Oxide | | Fluoride | |
|---|---|---|---|
| Host and Method | Covalency | Host and Method | Covalency |
| NiO (neutrons)[a] | $f_\sigma + f_s = 3.8\%$ | $KNiF_3$ (neutrons)[c] | $f_\sigma + f_s = 2.6 \pm 1.8\%$ |
| MgO (ENDOR)[b] | $f_\sigma = 8.5\%, f_s = 0.7\%$ | $KNiF_3$ (NMR)[d] | $f_\sigma = 3.8\%\ f_s = 0.54\%$ |
| | | $KMgF_3$ (ESR)[e] | $f_\sigma = 3.1\%\ f_s = 0.53\%$ |

[a] Ref. (56).
[b] Ref. (77).
[c] Ref. (82).
[d] Ref. (83).
[e] Ref. (84).

Does the new value indicate that oxides are indeed significantly more covalent than fluorides even for $Ni^{2+}$, and if so, is there some error in the neutron diffraction results? Such a conclusion would be of general concern. But in spite of the uncertainty in the magnitude of the zero-point spin deviation there does not seem to be a large source of error in the experimental technique or interpretation. It is more likely that with the extra information now available a more flexible interpretation is possible. Initially, almost all the resonance data obtained were for fluorides[22] (magnetically concentrated if by NMR and dilute if by ESR — with

---

[21] $KNiF_3$ has simple cubic G-type ordering — see Fig. 22.
[22] $^{16}O$ has $I = 0$, and oxides cannot be studied unless doped with $^{17}O$.

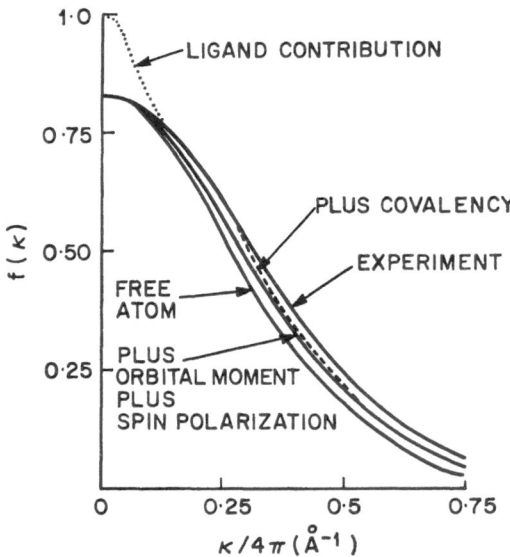

Fig. 21. Experimental and theoretical form factors for NiO [after Ref. (26)]. After inclusion of the effects of covalency, the orbital moment, and inner shell spin polarization, the calculated curve is in good agreement with the experimental figure [Ref. (27)], especially for $\varkappa/4\pi < 0.3$

similar results) and the first neutron data were almost all obtained for oxides — on concentrated samples. To compare data it was necessary to assume similar covalency for oxides and fluorides. But recently the acquisition of more data (particularly for $d^5$ ions) has indicated that, although this assumption does indeed appear to be valid for divalent salts, trivalent oxides are more covalent than trivalent fluorides and each is more covalent than the divalent isoelectronic salt (see below). It was still, however, necessary to combine neutron with resonance data to determine individual parameters ($f_\sigma$ and $f_\pi$) for $d^5$ ions. The relative simplicity of the $d^8$ situation shows that it cannot be assumed, at least for oxides, that the same covalency parameter will be measured in concentrated and dilute situations, even where the M — O distances are very similar as for NiO and MgO. Note also the different values of $f_\sigma - f_\pi$ found for $Mn^{2+}$ in $CsMnCl_3$ and doped in $K_4CdCl_6$ (Section 2.2).

It seems experimentally clear that the covalency of a transition metal to fluorine bond does not vary significantly from host to host, unless the M — F distance is significantly changed (84), but for combination with the more polarizable oxide ion it is possible that $Ni^{2+}$ for example may show a greater covalent interaction with $O^{2-}$ than does $Mg^{2+}$ (in line with the electron affinities; $Ni^{2+}$: 18.15 eV, $Mg^{2+}$: 15.03 eV). An oxide ion surrounded by five $Mg^{2+}$ cations and one $Ni^{2+}$ in $Ni^{2+}$-doped MgO may be capable of a greater covalent interaction with the $Ni^{2+}$ dopant than will an oxide ion in NiO surrounded by six $Ni^{2+}$ ions. It is not clear how to adjust the dilute result to compare with the concentrated situation. Crudely, we may use the simple MO theory [Eq. (2.23)] and equate the charge transferred away from an oxide ion in the two cases. For NiO this is $6f_\sigma$ (neglecting $f_s$) or $\approx 0.2\ e$. For $Ni^{2+}$-doped MgO the charge transfer to the nickel atom is $\approx 0.09e$ by $\sigma$-bonding leaving $\approx 0.1$ electron to be transferred to five $Mg^{2+}$ atoms.

If this interpretation is realistic, it would mean that experiments on dilute systems, unless chosen very carefully, might not be very significant in discussing the properties of concentrated salts. Obviously, NMR data on $^{17}O$-doped concentrated oxides would shed more light on this problem (although this will probably not be very easy to obtain for NiO) as would comparative data for other more polarizable ligands such as $Cl^-$. Also, it will be interesting to investigate covalency variations as coordination number or structure is changed — e.g., in a perovskite such as $KNiF_3$ each anion has only two $Ni^{2+}$ nearest neighbors, in $NiF_2$ three and in NiO six. If the total charge transferred to the metal remains the same as the coordination number of the anions is changed, the spin density associated with each anion will change. Such effects are widely accepted to occur when the coordination number of the cation changes (e.g., from octahedral to tetrahedral).

Although much theoretical and experimental work has gone into the study of bonding in transition metal salts the above discussion indicates the limited perspective that can be obtained even now from the existing spin density data. Many aspects remain to be understood, and hopefully the advances in neutron scattering techniques discussed in Section 3 will allow a rapid increase in the amount of data collected on spin distributions for different ions in various environments. Although the equivalence of 'concentrated' and 'dilute' data will be assumed in the discussion of $d^3$ and $d^5$ ions below, the questions raised by the uncertainty for $Ni^{2+}$ in oxide coordination should be borne in mind.

Although there has been no neutron investigation of $Ni^{3+}$, this ion has recently been studied in $^{17}O$-doped MgO (85). $Ni^{3+}$ is low spin ($t_{2g}^6 e_g^1$) in this situation and the EPR spectrum at 4.2K is characteristic of a dynamic Jahn-Teller effect. At 77K $^{17}O$ hyperfine structure could be measured and $f_s$ and $f_\sigma$ [defined by Eq. (2.21)] were estimated to be 0.83% and 9.4% respectively. *Freund* (77) suggested that these values, similar to those obtained for $Ni^{2+}$ in MgO indicated similar covalency for the two valence states of nickel. With one less $\sigma$-antibonding electron, however, the net ligand-to-metal charge transfer will be significantly greater for $Ni^{3+}$, as expected, even if the spin transfers are similar. The iso-electronic high spin ($t_{2g}^5 e_g^2$) $d^7$ ions $Co^{2+}$ and $Ni^{3+}$ were investigated in $KMgF_3$ some time ago by *Hall et al.* (84). Only an average of $f_\sigma$ and $f_\pi$ could be determined from the LHFI data but a significant increase in going from the divalent to the trivalent ion was found (2.4% and 8.6% respectively), in line with the data for $Ni^{2+}$ and $Ni^{3+}$ and for the $d^5$ ions $Mn^{2+}$ and $Fe^{3+}$.

No neutron data have been obtained for $Ni^{2+}$ for anions other than $F^-$ or $O^{2-}$ but NMR of $CsNiCl_3$ (86) gave $f_s = 0.58\%$ and $f_\sigma - f_\pi = 7.3\%$, a significant increase relative to $F^-$. This is in accord with the increase in covalency from MnO to MnS determined by neutron diffraction (56) and the ESR data for $Co^{2+}$ halides [$\frac{1}{2}(f_\sigma + f_\pi) = 2.4\%$ (87), 5% (88), 5.3% (89) and 7.5% (89) for doping in $KMgF_3$, $CdCl_2$, $CdBr_2$ and $CdI_2$, respectively].

The computation of spin transfer coefficients ($f$'s) and ligand field splittings ($\Delta$) has been pursued particularly actively for $NiF_6^{4-}$, starting with the pioneering work of *Sugano* and *Shulman* (90). *Wachters* and *Nieuwpoort* (91) discuss the dozen or so previous first principles calculations on this system indicating the various approximations used, and present a restricted Hartree-Fock self-consistent-field MO calculation for $NiF_6^{4-}$ in vacuo and in $KNiF_3$. Spin polarization is not very significant in affecting the properties of $d^8$ ions and these results are probably representative of the current computational situation, in which all electrons are included. Values of $f_s$ and $f_\sigma$ of 0.46% and 2.86% were obtained, in quite good agreement with experiment, and $\Delta$ was calculated to be 5440 cm$^{-1}$ cf. the experimental value of 7250 cm$^{-1}$ (90). The calculation did not take into account the effect of configuration interaction with charge transfer states which may increase $\Delta$. These authors argue that the largest contribution to $\Delta$ is in fact the overlap repulsion between metal and ligand electrons, an ionic term (although some other calculations mentioned in Ref. (91) find a higher covalent contribution) and find that all the $\sigma$-bonding MO's, except $3d_\sigma$, are expanded relative to the free ion orbitals. The $3d_\pi$ orbitals are more expanded but the $p\pi$ orbitals more contracted than the free-ion orbitals. The origin of the lowering of the spin-orbit parameters observed for transition metal complexes (11), and the reduced Coulombic repulsions (23) is thus still open to question (4). The reduction in the Racah parameters $B$ and $C$ is calculated to be about 2% compared to 10% experimentally. It is suggested that the effect of charge-transfer states on the "nephelauxetic effect" may be significant [see also Ref. (92)]. Spin unrestricted Hartree-Fock calculations of *Brown* and *Burton* (93) give quite good estimates of the effects of spin polarization for $Cr^{3+}$ but not for $NiF_6^{4-}$ (for reasons discussed by the authors). Spin density data for several complexes have also recently been calculated (94) by the multiple scattering $X\alpha$ method (95). A value of $f_\sigma = 6.3\%$ was obtained for $NiF_6^{4-}$, somewhat higher than the experimental figure, as was the case for all the other ions investigated.

The band structures of the transition metal monoxides including NiO have been a topic of considerable interest for many years, and study of spectra and transport properties continues in an effort to determine band widths, separations and electrostatic correlation energies. NiO is a *Mott* insulator (96) and the localized electron description assumed here is probably appropriate. Augmented plane wave band structure calculations have recently been made for NiO and other monoxides (97) and a localized electron multiple scattering $X\alpha$ calculation for NiO (98). Neither type of calculation includes electron-electron correlation effects.

$Ni^{2+}$ compounds have been investigated more thoroughly by calculation than those of any other transition metal ion. Although much progress has been made in interpreting the experimental observations, there is clearly still much to be done, as is also the case for the experimental investigation of covalency.

## 4.2 $d^3$ Ions — $Cr^{3+}$, $Mn^{4+}$

For $d^3$ ions, as for $d^8$ ions, the simple MO model indicates that because spin density is transferred via only one type of orbital, neutron and LHFI measurements should give similar results. Resonance data should give $f_\pi$ directly and the moment reduction observed by neutrons should be $4f_\pi$.

Spin polarization effects, however, make the situation more complicated. In a system containing unpaired electrons, exchange coupling will lower the energy of an orbital with parallel spin relative to its antiparallel-spin pair. Thus there will be a difference in covalency and spin transfer between up and down spin orbitals (we take the unpaired electrons always to be up spin). The difference will be proportional to the number of unpaired electrons and the energy difference and overlap between the polarizing and polarized orbitals, and spin polarization effects appear to be particularly significant for the $e_g$ $\sigma$-bonding orbitals in $d^3$ ions, which are spin polarized by the three unpaired $t_{2g}$ $\pi$-antibonding electrons. The result is to leave a net down spin on the ligands in the $e_g$ orbitals which contributes a term $-f_\sigma$ to the LHFI and conversely to transfer a net up spin to the metal $d_\sigma$ orbitals which reduces the observed moment reduction. A smaller effect is observed for the $4s$ orbitals in $d^3$ systems via the $a_{1g}$ $\sigma$-bonding orbitals and a $4s$ contribution is thought to occur (36) also for $d^5$ systems (Section 4.3), although in this latter case it cannot be observed directly, because $\sigma$-bonding effects are present in the first place.

The small negative value of $f_s$ ($\approx -0.1\%$) observed in ESR and NMR measurements of the LHFI in fluorides first demonstrated the effect of spin polarization. A significantly larger value ($-2.6\%$) was observed (99) for the $^{13}C$ LHFI in $Cr(CN)_6^{3-}$ and no doubt reflects the greater $\sigma$ bonding in the cyanide which also gives rise to the large ligand field splitting. An ENDOR study (100) of $^{27}Al$ super-transferred hyperfine interactions (STHI) in $Cr^{3+}$-doped $LaAlO_3$ revealed an interaction for the linear Cr-O-Al situation, also only explicable in terms of spin polarization. The most dramatic demonstration of the $e_g$ spin polarization was however revealed in the neutron diffraction study (62) of the moment reductions in $LaCrO_3$ and $CaMnO_3$. Because $e_g$ spin polarization ($-f_\sigma$) increases the apparent spin transfer $f_\pi$ as measured by LHFI (to $-f_\sigma - f_\pi$) and decreases the apparent moment reduction $4f_\pi$ (to $4f_\pi - 2f_\sigma$), the apparent values of $f_\pi$ (without taking spin polarization into account) for these oxides obtained by neutron diffraction were less than those observed for $Cr^{3+}$ and $Mn^{4+}$ fluorides by measurement of the LHFI. In this work (62) it was necessary to assume similar covalency for fluorides and oxides to derive $f_\sigma$ and $f_\pi$ independently, but more recent investigation of $CrF_3$ by neutrons (101) and $Cr^{3+}$ in $^{17}O$-doped MgO by ENDOR (102) have allowed the separate determination of $f_\sigma$ and $f_\pi$ for $Cr^{3+}$ in fluoride and in oxide coordination.

The data for $Cr^{3+}$ and $Mn^{4+}$ are given in Tables 4 and 5. There has been no neutron diffraction study of $V^{2+}$ [23]), but LHFI was observed (103) by ENDOR for $V^{2+}$ in $KMgF_3$ and the value of $f_\sigma - f_\pi$ ($-2.9 \pm 0.1\%$) indicates lower covalency than for $Cr^{3+}$ (Table 4). The observed $f_s$, however, is larger for $V^{2+}$ ($-0.10 \pm 0.01\%$) than for $Cr^{3+}$ in $KMgF_3$ ($-0.03\%$) and $Mn^{4+}$ in $Cs_2GeF_6$ ($+0.01 \pm 0.03\%$). This possibly (103) reflects either the increasing $3d-4s$ splitting with increasing oxidation state or an increasing effect of core polarization due to the unpaired spin in the ligand $2p$ orbitals and which gives a positive contribution to $f_s$.

---

[23]) Because of the relatively large radial extent of the $3d$ wavefunctions for lower-valent ions at the beginning of the transition series (leading to greater metal-ligand overlap) and the relatively small $3d-4s$ splitting, collective electron behavior is often observed for concentrated compounds of such ions — thus 'TiO' is metallic and 'VO' does not show magnetic ordering.

Table 4. Spin transfer coefficients for $Cr^{3+}$ in fluoride and oxide lattices

| | Fluorides | | Oxides |
|---|---|---|---|
| $f_s$ | $-0.031 \pm 0.004\%$ (in $KMgF_3$, ESR)[a] | $f_s$ | $-0.14 \pm 0.03\%$ (in MgO, ENDOR)[e] |
| | $-0.021\%$ (in $K_2NaGaF_6$, ESR)[b] | | |
| | $-0.021\%$ (in $K_2NaCrF_6$, NMR)[c] | | |
| [h]$f_\pi - f_\sigma$ | $4.9 \pm 0.2\%$ (in $KMgF_3$, ESR)[a] | [h]$f_\pi - f_\sigma$ | $7.1 \pm 0.7\%$ (in MgO, ENDOR)[e] |
| | $5.3\%$ (in $K_2NaGaF_6$, ESR)[b] | | |
| | $4.9\%$ (in $K_2NaCrF_6$, NMR)[c] | | |
| [h]$f_\pi + \frac{1}{2}f_\sigma + \frac{1}{2}f_s$ | $1.5 \pm 0.4\%$[i] (in $CrF_3$, neutrons)[d] | [h]$f_\pi + \frac{1}{2}f_\sigma + \frac{1}{2}f_s$ | $3.7\%$ (in $LaCrO_3$, neutrons)[f] |
| | | | $2.2 \pm 0.6\%$[i] (in $LaCrO_3$, neutrons)[g] |
| [h]$f_\sigma = -2.3 \pm 0.4\%$ | | [h]$f_\sigma = -3.2 \pm 0.9\%$ | |
| $f_\pi = 2.6 \pm 0.4\%$ | | $f_\pi = 3.9 \pm 0.6\%$ | |

[a]) Ref. (*84*).
[b]) Ref. (*104*).
[c]) Ref. (*105*).
[d]) Ref. (*107*)[i].
[e]) Ref. (*102*).
[f]) Ref. (*64*).
[g]) Ref. (*62*)[i].
[h]) In the absence of spin polarization $f_\sigma$ would be zero.
[i]) These values have been corrected to the revised scattering length of germanium [Ref. (*49*)].


Table 5. Spin transfer coefficients for $Mn^{4+}$ in fluoride and oxide lattices

| | |
|---|---|
| $f_s$ | $0.10\%$ (in $Cs_2GeF_6$, ESR)[a] |
| [c]$f_\pi - f_\sigma$ | $9.2\%$ (in $Cs_2GeF_6$, ESR)[a] |
| [c]$f_\pi + \frac{1}{2}f_\sigma + \frac{1}{2}f_s$ | $3.8 \pm 0.8\%$ (in $CaMnO_3$, neutrons)[b] |
| [c]$f_\sigma = -3.6 \pm 0.8\%$[d] | |
| $f_\pi = 5.6 \pm 0.8\%$[d] | |

[a]) Ref. (*104*).
[b]) Ref. (*62*) [This value has been corrected to the revised scattering length of germanium [Ref. (*49*)]].
[c]) In the absence of spin polarization $f_\sigma$ would be zero.
[d]) Values derived assuming similar covalency for oxides and fluorides.

The moment reduction in LaCrO$_3$ was first studied by *Nathans et al.* (*64*) who measured $f_\pi$ to be 4.6%[24]), assuming no spin polarization, compared with 2.2 ±0.6% found in the later determination of *Tofield* and *Fender* (*62*). LaCrO$_3$ (in common with CaMnO$_3$ and LaFeO$_3$) is an orthorhombic perovskite and has G-type magnetic ordering (Fig. 22) reflecting the 180° superexchange interactions but the nuclear structure was not accurately known at the time of the neutron diffraction studies. MnO was used as an external calibrant in the first determination [(011 + 101)$_{mag}$ cf. (111)$_{mag}$ of MnO] and Ge in the second [(011 + 101)$_{mag}$ cf. (111)$_{Ge}$]. In the latter work considerable care was taken to eliminate as many sources of error as possible and the result is considered reliable. An external calibration is of course sensitive to errors in the scattering length of the calibrant and the data from Refs. (*62*) and (*101*) have been corrected to be consistent with a revised value (*49*) of the germanium scattering length.

The thorough investigation of FeF$_3$ (*101*) by profile analysis, by internal calibration [(111 + 100)$_{mag}$ cf. (110)$_{nucl}$ — rhombohedral indexing] and by

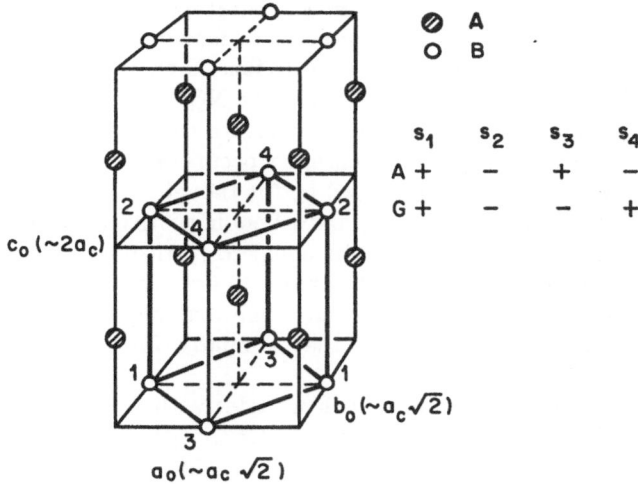

Fig. 22. Magnetic ordering in simple perovskites. The orthorhombic chemical unit cell found for LaMO$_3$ (M = Mn, Cr, Fe) and YFeO$_3$ is indicated and the relationship of the lattice parameters to the pseudocubic subcell shown (KNiF$_3$ is cubic and the chemical cell is then the smaller primitive perovskite cube indicated by heavy lines). The orthorhombic magnetic unit cell is the same size as the chemical unit cell. KNiF$_3$, LaFeO$_3$, LaCrO$_3$, CaMnO$_3$ and YFeO$_3$ have G-type ordering and LaMnO$_3$ has A-type ordering. The sites of the 12-fold coordinate metal atoms (A) and the octahedrally coordinated transition metal atoms (B) are shown but the oxygen atoms are omitted. For KNiF$_3$ the magnetic unit cell can be described by a cubic cell, twice the size of the chemical cell and the lowest angle magnetic reflection is (111) based on the doubled cell. For the orthorhombic G-type perovskites this is split into two reflections [orthorhombic (011) and (101)], the relative intensity of which is determined by the orientation of the spins

---

24) Which is reduced to 3.7% if the zero-point spin correction of *Davis* (*67*) used by *Tofield* and *Fender* (*62*) is invoked.

external calibration with germanium [$(111 + 100)_{mag}$ cf. $(111)_{Ge}$] gave a somewhat higher value of $f_\sigma + 2f_\pi + f_s$ for external calibration (9.0 $\pm$ 0.7% after correction for revised Ge scattering length) compared with those obtained by internal calibration (5.5 $\pm$ 0.8%) and by profile analysis (5.9 $\pm$ 0.7%). Nevertheless, these values are in qualitative agreement with regard to the conclusions to be drawn and the data for $Fe^{3+} - O^{2-}$ obtained by germanium calibration (62) are in reasonable agreement with other values measured by profile analysis (Section 4.3). Also, the data for $Cr^{3+} - F^-$ (101) agree with the conclusions of *Tofield* and *Fender* (62) for LaCrO$_3$, and this latter determination is used in the analysis of $f_\sigma$ and $f_\pi$ presented here.

The data of Table 4 indicate an increase in both ligand-to-metal $\pi$-bonding and in spin polarization in going from fluoride to oxide coordination. This increase in covalency when bonding to the more polarizable oxide ion is not unexpected and agrees with the trend found for $Fe^{3+}$, although the data for $Ni^{2+}$ and $Mn^{2+}$ are possibly indicative of similar covalency for divalent oxides and fluorides. A comparison of $f_\pi$ and spin polarized $f_\sigma$ indicates that $\sigma$ and $\pi$-bonding are possibly of similar magnitude for the $d^3$ ions, in contrast to the situation found for $Fe^{3+}$ where $\sigma$-bonding is dominant. Also, from Table 5 it would appear that $\pi$-bonding increases from trivalent chromium to tetravalent manganese. The greater importance of $\pi$-bonding for $d^3$ ions relative to $d^5$ ions is an interesting observation. Strong $\pi$-bonding effects are also observed in the NQR spectra of early transition metal complexes for which a positive temperature coefficient of the quadrupole coupling constant is often observed. This effect occurs only for complexes with no, or few antibonding $\pi$ electrons ($d^0$ to $d^3$) and is thought (109) to be the result of bending vibrations increasing $\pi$-overlap and $\pi$-bonding with increasing temperature.

If an average $f_\sigma$[25]) is assumed to be equal to $f_\pi$ then via Eq. (2.23) charge transfers of $\sim$0.6e and $\sim$0.9e from the anions to $Cr^{3+}$ are found for fluoride and oxide coordination respectively. These are somewhat higher than estimated for $Fe^{3+}$ but the oxide value is similar to that estimated for low-spin $Ni^{3+}$ in MgO (Section 4.1).

A number of ternary chromium chalcogenides with fairly complicated magnetic structures have recently been investigated by powder neutron diffraction using profile analysis, and effective moments of 2.36, 2.55, 2.48, 2.26 and 3.04 were found for NaCrSe$_2$, AgCrSe$_2$, NaCrS$_2$ (106), LiCrS$_2$ (107) and KCrS$_2$ (108) respectively. These data indicate an uncorrected $f_\pi$ of zero for KCrS$_2$ up to 5.3% for LiCrS$_2$, [using the same zero point spin correction (4.2%) (67) as for LaCrO$_3$ and CrF$_3$]. Although it would be interesting to have data for more polarizable ligands such as $S^{2-}$ and $Se^{2-}$ bonded to $Cr^{3+}$, it is probably premature to draw any conclusions about the magnitude of the spin reductions to be expected.

Only one form factor determination has been made for a $d^3$ ion. $Cr^{3+}$ in paramagnetic K$_2$NaCrF$_6$, magnetically aligned at 4.2 K, was studied using polar-

---

[25]) $f_\sigma$ is of course zero in the absence of spin polarization .The value assigned is a measure of $\lambda_\sigma$ [Eq. (2.21)].

ized neutrons (*22*). This crystal contains isolated $CrF_6^{3-}$ octahedra and the data may be unambiguously interpreted as arising from the unpaired spin density of isolated clusters. The structure is simple, having only one variable positional parameter for the fluoride ion, and an accurate structure analysis was carried out at 4.2 K to provide the set of nuclear structure factors. The experimental form factor was normalized to unity at $\varkappa = 0$ by measuring the bulk magnetization at the same field and temperature as in the polarized neutron experiment. The deviation from spherical symmetry of the $t_{2g}^3$ electron configuration is reflected in the measured form factors along different symmetry directions. The form factors along [h00] and [hhh] are shown in Fig. 23. Broken lines are sketched through the data and the full lines are calculated free ion functions (*79*).

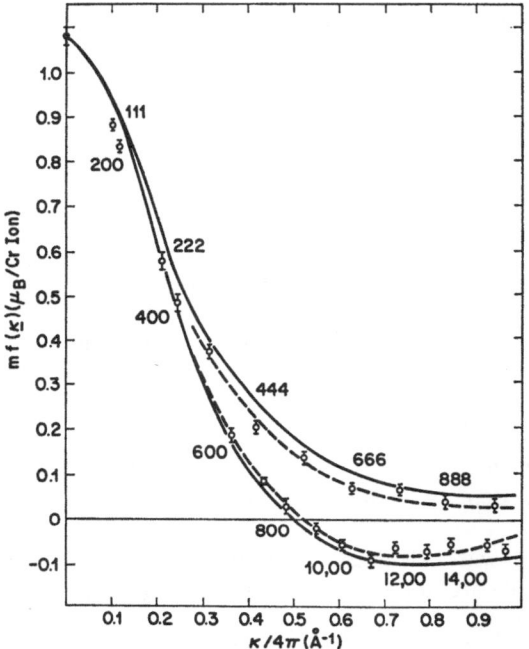

Fig. 23. Magnetic scattering amplitudes $mf(\varkappa)$ per chromium atom in $K_2NaCrF_6$ at 4.2 K and 17.6 kOe along the [h00] and [hhh] symmetry directions. Full lines are theoretical free ion functions. Broken lines are sketched through the data [after Ref. (*22*)]

The spin density (Fig. 24) determined by Fourier transformation at 0.4 Å resolution of the magnetic structure factor data clearly shows the $t_{2g}$ nature of the spin density on $Cr^{3+}$. Also apparent is spin density covalently transferred to the fluorides. This is not exactly centered on the $F^-$ sites because the antibonding nature of the wavefunctions with unpaired spin gives rise to the negative overlap region between the metal and ligand ions [Sections 2.2 and Eq. (2.28)] which pushes out the maximum of the ligand spin density. Because of the finite resolution of the map the nodes along the *x* and *y* axes of the chromium are washed out and the

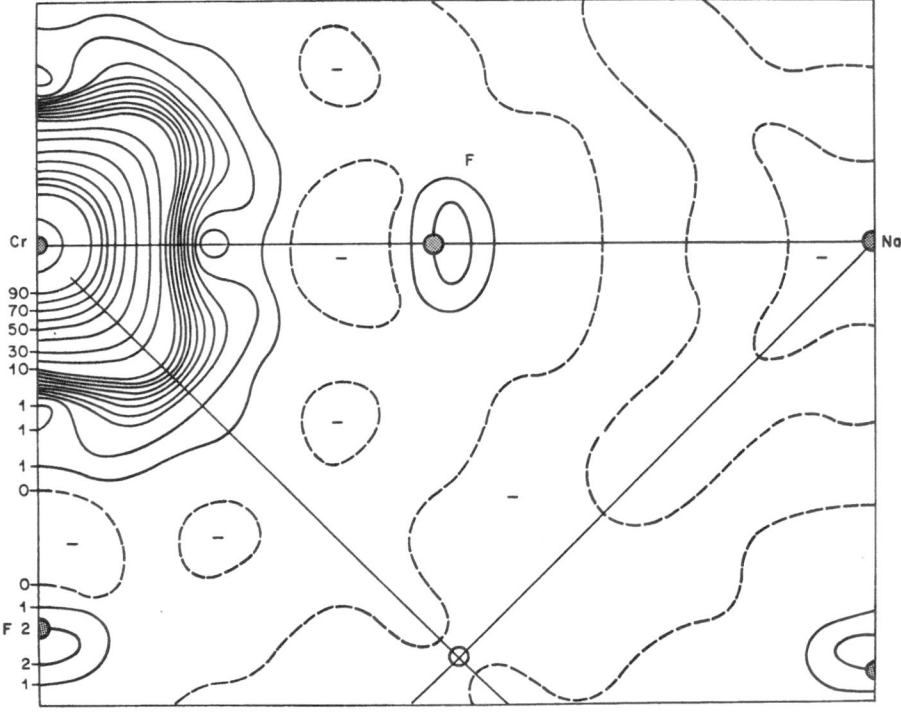

Fig. 24. Fourier transform of the magnetic scattering amplitudes of $K_2NaCrF_6$ in the (001) plane through a chromium site. Contour units are in 0.01 $\mu_B/\text{Å}^3$ [after Ref. (22)]

shape of the spin density distribution on the fluorines cannot be precisely seen. The majority spin density would appear to be associated with the $p_\pi$ orbital, however.

A form factor was calculated from free ion $Cr^{3+}$ and $F^-$ wave functions assuming the simple MO model. From Eqs. (2.8)–(2.10) and (2.24) this is

$$f(\varkappa) = N_\pi^2 \{f_{dd}(\varkappa) + \lambda_\pi^2 f_{pp}(\varkappa) + 4\lambda_\pi S_\pi f_{pd}(\varkappa)\} \tag{4.1}$$

where $f_{dd}(\varkappa)$ and $f_{pp}(\varkappa)$ are metal only and ligand only form factors and $f_{pd}(\varkappa)$ is an overlap form factor. $S_\pi = \langle d|p \rangle$. *Wedgwood* (22) expanded the three separate contributions to $f(\varkappa)$ in terms of cubic harmonics and determined the contribution of each to the spherical zero-order and aspherical fourth-order form factors $f_0(\varkappa)$ and $f_4(\varkappa)$ for $f_\pi = 0$, for $f_\pi = 5.8\%$ ($\lambda_\pi = 0.48$), close to the value obtained (105) by *Shulman* and *Knox* by NMR for this material (Table 4), and for $f_\pi = 2.3\%$ ($\lambda_\pi = 0.30$), close to the value found by neutron diffraction for $LaCrO_3$ and $CrF_3$ (Table 4). For comparison, the zero-order and fourth-order experimental form factors were extracted from the experimental $f(\varkappa)$ by a double Fourier transform method (110). The comparison is shown in Figs. 25 and 26. These figures indicate deficiencies both in the simple MO model and in the Fourier transform procedure (because of termination errors). Although there is signifi-

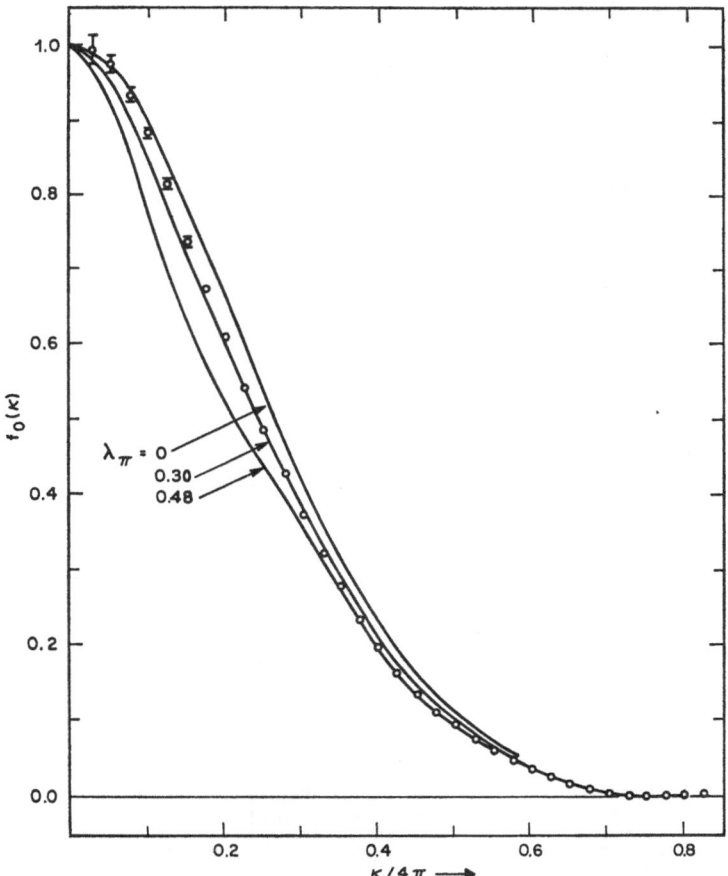

Fig. 25. Experimental values of $f_0(\varkappa)$ (circles) and $f_0(\varkappa)$ given by molecular-orbital theory (full lines) for increasing covalency ($\lambda_\pi = 0$, 0.30 and 0.48) [after Ref. (22)]

cant uncertainty in such a fitting procedure, the lower value of $\lambda_\pi$ (0.30) seems to give the best fit to the spherical form factor data, but for $\varkappa/4 \leq 0.4$ the asherical form factor agrees well with $\lambda_\pi = 0.48$. This difference may well reflect the existence of $e_g$ spin polarization already discussed and which was not included in the MO model. The neutron experiment determines the complete spin density, but $f_4(\varkappa)$, and particularly the positive hump at low $\varkappa$, is primarily sensitive to the ligand spin and should agree with the LHFI determination, whereas $f_0(\varkappa)$ is primarily sensitive to the metal spin except at low $\varkappa$. The oscillation in $f_4(\varkappa)$ for $\varkappa/4\pi > 0.4$ is the result of termination errors in the Fourier procedure and in fact oscillations in both $f_0(\varkappa)$ and $f_4(\varkappa)$ which are greater than the effects of covalency are found in this region. However, *Wedgwood* pointed out that an effect of $e_g$ spin polarization would be to reduce the magnitude of $f_4(\varkappa)$ at large $\varkappa$. Although the Fourier oscillations mask any effect, $f_4(\varkappa)$ may be determined directly from the experimental $f(\varkappa)$ along different directions in the crystal (dashed lines, Fig. 23) and this estimate, drawn as a dashed line in Fig. 26, shows the predicted effect.

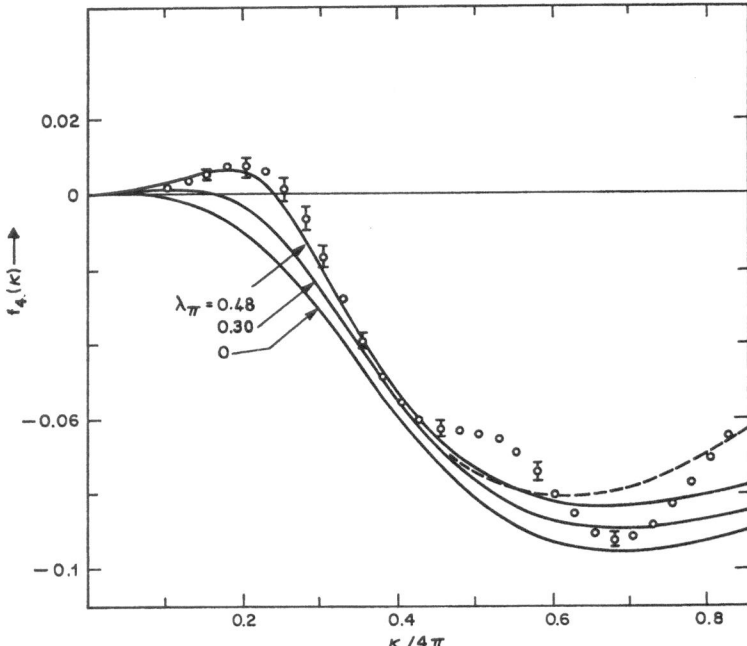

Fig. 26. Experimental values of $f_4(\varkappa)$ (circles) and $f_4(\varkappa)$ given by molecular-orbital theory (full lines) for increasing covalency ($\lambda_\pi = 0$, 0.30 and 0.48) [after Ref. (22)]

Such a form factor determination provides the most stringent test for any theoretical description of a transition metal complex. Unfortunately, the experiment described is a representative of only one or two such measurements determined sufficiently accurately to provide such a test. This particular experiment was significant, both in demonstrating the shape of the $d^3$ $t_{2g}$ magnetic electron distribution for the first time, and in indicating the particular sensitivity of the aspherical form factor component $f_4(\varkappa)$ to the covalent terms in the spin distribution. The shape of $f_4(\varkappa)$ strongly supported the inference for $e_g$ spin polarization drawn (62) from the comparison of powder neutron diffraction data and resonance data. Measurement of the aspherical form factor can, in principle, be performed relatively easily without incurring such severe problems of extinction as accompany the measurement of the total spin density distribution using single crystals. It is likely that such measurements will soon begin to provide much detailed information concerning covalent effects on spin distributions both with single crystal, and, hopefully, with polycrystalline samples.

There have been few first principles calculations for $d^3$ ions and no form factor calculations. Brown and Burton's spin-unrestricted Hartree-Fock SCFMO calculations (93) for $K_2NaCrF_6$ gave $f_\sigma = -2.19\%$, $f_\pi = 2.55\%$ and $f_\sigma - f_\pi = -4.74\%$, in quite good agreement with the experimental estimates (Table 4), but a somewhat small value of $f_\sigma - f_\pi$ ($\sim -6.5\%$) was estimated for $MnF_6^{2-}$ (cf. Table 5). On the other hand the $X\alpha$ multiple scattering calculation of *Larsson* and *Connolly* (94) indicates that the contribution to $f_\sigma - f_\pi$ from $e_g$ spin polarization is dominant for $CrF_6^{3-}$, $MnF_6^{2-}$ and $CrCl_6^{3-}$ (for $CrF_6^{3-}$, $f_\sigma = -4.8\%$, $f_\pi = 1.0\%$, $f_\sigma - f_\pi = -5.8\%$). This conclusion would seem to be difficult to reconcile with the neutron scattering data. For $CrCl_6^{3-}$ they find $f_\sigma = -10.2\%$, $f_\pi = 0.3\%$, and $f_\sigma - f_\pi = -10.5\%$. Such a large spin polarization effect should be easily detectable in an accurate form factor measurement or in a combination of $f_\sigma$ and $f_\pi$ determined by LHFI and powder neutron diffraction.

Although the presence of spin transferred to the ligands by covalency effects has been observed in several polarized neutron experiments, and $f_4(\varkappa)$ determined for $K_2NaCrF_6$ showed the presence of the ligand moment, the 'forward peak' in $f_0(\varkappa)$ [Eq. (2.27) and Fig. 5], predicted to occur at $\varkappa/4\pi \lesssim 0.1$ by *Hubbard* and *Marshall* (26), and which should provide a fairly direct measure of the covalency parameter sum has proved very difficult to observe. It is quenched in most antiferromagnets and in very few other magnetically ordered systems are there magnetic reflections at sufficiently low angle to observe the effect by magnetic Bragg scattering measurements. Thus it appeared (17) that the simplest approach would be to investigate paramagnetic systems, where the metal-ligand clusters do not overlap, by long wavelength neutron diffuse scattering [Eqs. (3.23) — (3.25) and Section 3.8].

In order to avoid significant exchange coupling between the magnetic ions such systems are necessarily relatively dilute. This raises problems of intensity, particularly as in many powders low angle diffuse scattering, probably caused by surface effects or crystal defects, is observed (17) and which often has a much greater cross-section than expected for any paramagnetic scattering. In crystals such as $K_2NaCrF_6$ however, where the magnetic ions are an integral part of the structure and their concentration may be as high as 10 atomic percent, but where the interactions between magnetic ions are extremely weak so that ordering temperatures are no greater than $\sim 1K$, the paramagnetic cross-section is sufficiently large to be measurable by a magnetic switching experiment [Eq. (3.25), Fig. 7]. Polycrystalline $K_2NaCrF_6$ was investigated at 4.2 K (17) but because of the regular arrangement of the $Cr^{3+}$ ions, multiple Bragg scattering was observed with the field on in the $\varkappa$ region of interest, and obscured the paramagnetic effects. It is possible that this drawback could be overcome by suitable orientation of a single crystal. Of course, for this type of crystal with a somewhat larger unit cell, the low angle region could be directly investigated by Bragg scattering with the polarized beam.

A qualitative observation of the ligand forward peak was made for $Cr^{3+}$ in single-crystal $Al_2O_3$ (ruby) (20). A total diffuse scattering cross-section measurement was made at room temperature on a ruby crystal of length 2 cm and diameter 1 cm containing 1.26% (atomic) of chromium and also on a pure sapphire ($Al_2O_3$) single crystal of the same dimensions (Fig. 27). The isotropic scattering from the sapphire indicated that multiple Bragg scattering was not significant. By subtracting the sapphire cross-section (0.4 mbarn) the incoherent scattering of oxygen and aluminum were accounted for as well as the effects of thermal diffuse scattering and any multiple scattering. Inelastic effects between exchange coupled $Cr^{3+}$ ions were estimated to have only a small influence on the expected paramagnetic scattering and the incoherent disorder scattering arising from the distribution of aluminum and chromium on the cation sublattice [Eq. (3.11)] was by chance

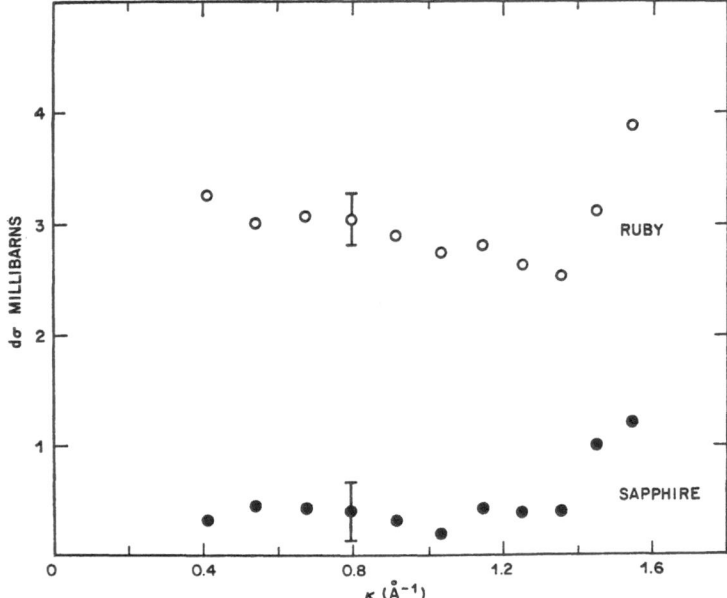

Fig. 27. Measured total diffuse differential scattering cross-section of single crystal sapphire and ruby (1.26% atomic Cr). $\lambda = 6.0$Å. The statistical error is indicated. Total counting time was 48 hours [after Ref. (17)]

very small because of their very similar scattering lengths. The ruby-sapphire difference cross-section thus had contributions only from the chromium paramagnetic scattering and the chromium incoherent scattering [Eq. (3.9)]. Incoherent cross-sections are in general not at all accurately known, but the difference cross-section was anyway somewhat smaller than expected, even for $\sigma_i^{Cr} = 0$. This was probably the result of calibration errors because of the very small cross-sections (20) involved (the background scattering was significantly greater than the paramagnetic scattering). The form factor was therefore calibrated by assuming $f(\varkappa) = 0.98$ at the lowest angle measured. The resulting form factor, together with free-ion curves (79) normalized to 1.0 and 0.9 are shown in Fig. 28. The errors (mainly statistical) of individual points are large but the trend in values as exemplified by the linear plot is as expected for ligand paramagnetic scattering with $f(\varkappa)$ dropping much more rapidly for $0 < \varkappa < 1.3$ than the free ion curve. The data are consistent with a ligand moment of approximately 10%.

Until much improved flux-to-background ratios are available it does not seem likely that much progress will be made with this type of experiment either by observation of total diffuse cross-sections, magnetic-switching cross-sections, or polarization analysis measurement of paramagnetic cross-sections, and conventional form factor measurement of paramagnetic as well as of magnetically ordered systems will be much more fruitful.

One other determination of the ligand forward peak has recently been reported (41) however, in a somewhat different situation. $K_2CuF_4$ has the $K_2NiF_4$ structure

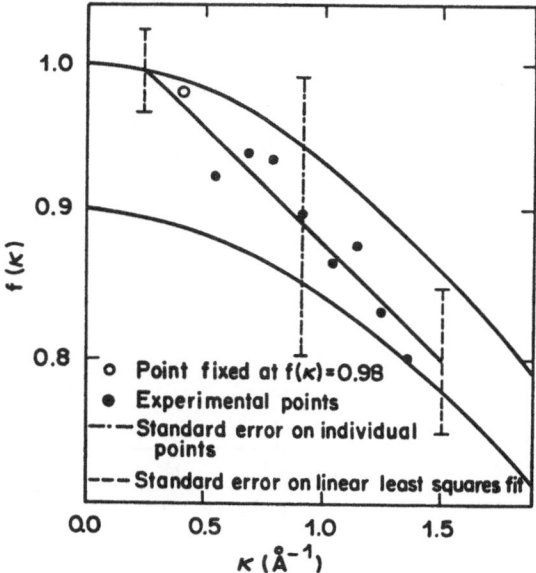

Fig. 28. Experimental form factor for $Cr^{3+}$ in $Al_2O_3$ [after Ref. (20)]. The full lines are theoretical free ion curves normalized by 1.0 and 0.9

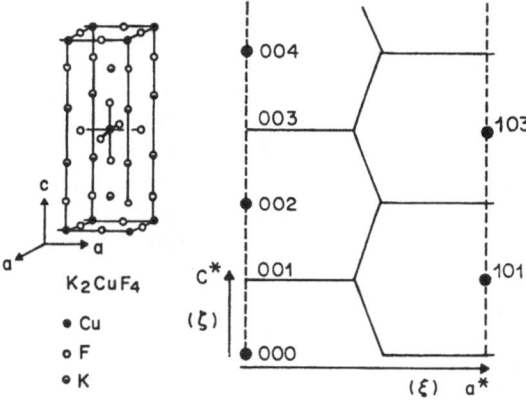

Fig. 29. The crystal structure of $K_2CuF_4$ and the $\xi 0\zeta$ section of the reciprocal lattice. The critical scattering ridge appears along the broken lines.
[After *Hirakawa K., Ikeda, H.*: J. Phys. Soc. Japan *35*, 1328 (1973)]

(with a slight orthorhombic distortion) with widely separated layers of somewhat distorted $CuF_6^{4-}$ octahedra (Fig. 29). Unlike $K_2NiF_4$ which is a two-dimensional antiferromagnet, $K_2CuF_4$ is a two-dimensional ferromagnet ($T_c = 6.25\,K$) and this property leads to pronounced critical scattering along the $00\zeta$ reciprocal lattice line (Fig. 29) where $\zeta$ is the coordinate variable along the $c^*$ direction (the octahedron is compressed along the $c$-axis leaving the unpaired spin in the $d_{3z^2-r^2}$ orbital).

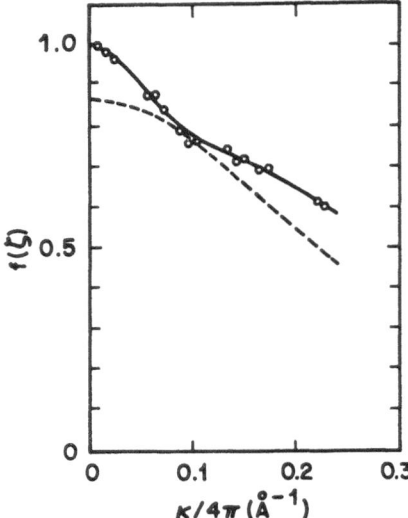

Fig. 30. The form factor for $Cu^{2+}$ in $K_2CuF_4$ [after Ref. (*41*)]. The broken curve is the theoretical form factor for isolated $Cu^{2+}$ taking into account the asphericity in the $00\zeta$ direction. This curve is renormalized so as to give the reduction factor given by the NMR data for $f_\sigma$ and $f_s$

The intensity of the critical scattering ridge is a function of $f(\varkappa)$ [$f(\zeta)$ in this case] and the form factor could be determined for $\zeta = 0.2$ up to 5.8 with reasonable accuracy (Fig. 30). Because of the ferromagnetic ordering the ligand spins are conserved and a foreward peak is indeed observed. The dashed line is a spin only free ion form factor normalized to 0.876[26], and the neutron result seems in reasonable agreement with NMR data (*111*). The form factor at $\varkappa/4\pi = 0.157$ (0.71) agrees quite well with the value (0.75) determined by measurement of the (004) magnetic Bragg reflection. Unfortunately, this type of measurement can only be performed in the rather unusual set of circumstances provided by the properties of $K_2CuF_4$.

The crystal field analogue of an octahedral $d^3$ ion is a tetrahedral $d^7$ ion and tetrahedral $Co^{2+}$ is fairly often encountered in crystals and complexes. Little measurement of covalency effects has been made, but several $Co^{2+}$-containing spinels have been investigated by neutron diffraction. The different magnetic interactions within and between the tetrahedral A sites and the octahedral B sites often lead to complex magnetic structures, and the frequently found occurrence of similar metal atoms in both A and B sites is another complication. $Co_3O_4$, however, is a simple normal spinel [$Co^{2+}$ in the A sites and diamagnetic low-spin ($t_{2g}^6$) $Co^{3+}$ in the B sites] with simple antiferromagnetic ordering (*112*) (nearest-neighbor sites being antiparallel). The investigation of *Roth* (*112*) did not, however, reveal any moment reduction for $Co^{2+}$ which was rather surprising.

---

[26] This reduction corresponds to the spin transfer to the two fluorines along the $c$-axis determined by NMR (*111*) [$2(f_\sigma + f_s) = 2(5.76 + 0.43)\% = 0.124$].

$Co_3O_4$ and the rhodium-containing analogue $CoRh_2O_4$ have recently been reinvestigated (72) using the profile analysis technique. A slight expansion of the $Co^{2+}$ form factor relative to the free ion calculation (79) was indicated in both cases and closely similar values of $g\langle S \rangle$ [2.747(8) for $Co_3O_4$ ($T_N \simeq 40K$) and 2.78(8) for $CoRh_2O_4$ ($T_N = 27K$)] were found at 1.2 K. Using $g = 2.27$ (found for $Co^{2+}$/ZnO) and correcting for the rather large zero-point spin reduction expected for a tetrahedral environment, a spin reduction of 12% was found for $Co_3O_4$. The moment reduction is quite similar to that found for NiO, for example (Table 3), which implies a greater covalency per bond ($4/3f = 0.12$ or $f = 8.9\%$ where $f$ is the fractional spin transfer per ligand) for the tetrahedrally coordinated $Co^{2+}$. Similar results have also been found recently for tetrahedral $Mn^{2+}$ in $MnRh_2O_4$ examined with $CoRh_2O_4$ (72) and for tetrahedral $Fe^{3+}$ (71, 73)[27]. The error in the earlier investigation of $Co_3O_4$ appears to have been due primarily to use of an incorrect value of the cobalt scattering length.

## 4.3 $d^5$ Ions — $Mn^{2+}$, $Fe^{3+}$

The high spin $d^5$ ions $Mn^{2+}$ and $Fe^{3+}$ have been studied as thoroughly as any transition metal ions, but there is still disagreement over their behavior and the correct interpretation of it — especially so for $Mn^{2+}$. These problems arise in the first place from the fact that both $\sigma$ and $\pi$ covalency contribute to moment reductions and LHFI parameters (Table 1), and even in the simple MO model a combination of neutron and resonance data is needed to determine individual $\lambda_\sigma$ and $\lambda_\pi$. This interpretation is confused if $4s$ spin polarization effects are present (quite possible for the $d^5$ high spin configuration) as was suggested by *Hubbard et al.* (36). Secondly, it has been assumed by some authors that only $\sigma$ bonding can be significant and therefore $Mn^{2+}$ should behave as does $Ni^{2+}$. In fact ($f_\sigma - f_\pi$) observed by measurement of LHFI, and the moment reductions for $MnF_2$ and MnO are much lower than the values found for $Ni^{2+}$ salts so that $Mn^{2+}$ is considered to show 'anomalously low' covalency. The similarity of the covalency parameters determined for the divalent oxides and fluorides of $Mn^{2+}$ and of $Ni^{2+}$ is considered a related problem — even though this is no longer the case for the more covalent trivalent ions $Cr^{3+}$ and $Fe^{3+}$. Finally, a measurement (113) of the $Mn^{2+}$ form factor in several polycrystalline compounds revealed a contraction relative to the calculated free ion value (79), indicating expanded metal $3d$ orbitals. Although such an effect was not observed for $Ni^{2+}$ (27), this neutron evidence was apparently support for the concept of central-field covalency and the nephelauxetic effect (23), and was used (114) to support evidence from Mössbauer quadrupole splitting measurements on $Fe^{2+}$ complexes on the same point [see Ref. (4)].

We will discuss first the form factor measurements for $Mn^{2+}$ and $Fe^{3+}$. The experiment of *Hastings et al.* (113) was performed before the effects of covalency on magnetic scattering intensities were realized and no account was taken of

---

[27]) And in the analysis of dielectric properties (34), and in Mössbauer effect and NQR measurements on $Sn^{IV}$ and $Pb^{IV}$ compounds [see Ref. (4)].

the moment reductions due to covalency or of zero-point spin deviation. It appeared probable that these factors might be significant and it seemed to be necessary therefore, to remeasure the form factor, taking advantage also of improvements in apparatus and data refinement. Polycrystalline MnO was an attractive candidate for such an experiment (48) as the simple nuclear and magnetic (Fig. 20) structures meant that a good separation of magnetic and nuclear intensities could be made. Aspects of the data refinement and $\lambda/2$ corrections have been mentioned already (Section 3.6) as have absorption corrections (Section 3.1) and the temperature factors used (Section 3.2). To obtain an accurate scale factor at 4.2K from the nuclear intensities (essential in providing a correct calibration of the magnetic intensities) it was also necessary to accurately fix the Mn scattering length. It was not possible to do this by unconstrained refinement of room temperature data ($T_N = 120$K) because of the high correlation between the scale factor, the scattering length and the temperature factors. By using calculated temperature factors however (115) nuclear refinements were carried out at room temperature and 4.2K to give a sufficiently accurate scattering length ($b_{Mn} = -0.372 \pm 0.005 \times 10^{-14}$ m). The calculated manganese temperature factor at 4.2K was used in the refinement of the magnetic intensities and a recent calculation (65) of the zero-point spin deviation for MnO was used. The form factor was measured to $\varkappa/4\pi = 0.54$ (11 magnetic intensities). Contraction relative to the free ion form factor was indeed found if the observed $\langle S \rangle f(\varkappa)$ were normalized to $\langle S \rangle = 5/2$ but after inclusion of the zero-point spin deviation and normalization to the free-ion form factor for the lowest angle magnetic reflection, $f(\varkappa)$ for all the other reflections followed closely the free ion curve (Fig. 31). The calibration to the free ion form factor is thought to be a good approximation (see Section 2.2 and Fig. 5) and this was the case for the $Ni^{2+}$ form factor in

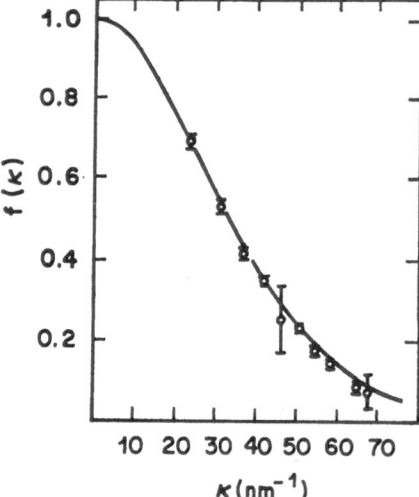

Fig. 31. The magnetic form factor for $Mn^{2+}$ in MnO [after Ref. (48)]. The experimental points have been normalized to the free ion form factor for the lowest angle reflection. The full line is the Hartree-Fock free ion curve

NiO (56) (Section 4.1). The only way to avoid such a calibration apart from extra-polation using a particular shape for $f(\varkappa)$ is to estimate the total spin by integration of the moment density determined by Fourier transformation of single crystal data, which carries its own problems (Section 3.7) as well as being a vastly more time-consuming experiment. This procedure has been carried out (73) however for octahedral and tetrahedral $Fe^{3+}$ in yttrium iron garnet (YIG).

Thus there appears neither to be experimental support for a contracted form factor as previously claimed, nor for an expanded form factor as in NiO. It is not possible to distinguish by experiment the effects of the overlap form factor (leading to expansion), radial expansion (contraction) or $4s$ polarization (con-traction). Two unrestricted all-electron SCF Hartree-Fock calculations have been made (116, 76) for the form factor associated with the $MnF_6^{4-}$ cluster. This was before the redetermination (48) of MnO was performed and the ambition was to explain the contracted form factor of *Hastings et al.* (113). *Freeman* and *Ellis* (116) indicated that expansion of the metal $t_{2g}$ $\pi$-bonding orbitals (but not of the $e_g$ $\sigma$-bonding orbitals) was significant. The calculation of *Soules* and *Richardson* (76) appears to be in closer agreement with the more recent data, and they indicate a negative contribution of $\sim 1\%$ to the form factor in the for-ward direction as the result of spin polarization. The suggestion of differential expansion of $e_g$ and $t_{2g}$ orbitals is interesting [although apparently in disagree-ment with a weak field interpretation of the ligand field spectra of $KMnF_3$ (92)]. A major contribution to the investigation of $Mn^{2+}$ would be a comprehensive form factor determination of the type attempted for $K_2NaCrF_6$.

Evidence for the inadequacy of the early $Mn^{2+}$ experimental form factor had also been revealed in an experiment on $MnCO_3$ by *Brown* and *Forsyth* (58), further analyzed by *Lindgard* and *Marshall* (117). $MnCO_3$ is basically antiferro-magnetic (Fig. 32) but spin canting (Section 3.6) produces a ferromagnetic moment of 0.036 $\mu_B$ perpendicular to the antiferromagnetic spin direction. Because of the crystal symmetry, the ferromagnetic component of the spin, but not the antiferromagnetic component, contributes to reflections $(hhl)$ with $l$ even, and the crystal was mounted so that these reflections could be measured with the polarized beam. Fourteen magnetic structure factors out to $\varkappa/4\pi = 0.42$ were recorded at fields of $H = 1.6$ kOe and 7 kOe in an attempt to distinguish between the effects of the spontaneous magnetization and the field-induced magnetiza-tion. Although the transformed spin density could only be obtained at low re-solution because of the lack of high angle data, a comparison of this density with a calculated spin density including covalent transfer on to the six nearest-neighbor oxygen atoms indicated the existence of spin polarization on the $CO_3^{2-}$ anion with a negative moment on the carbon atoms [the ligand density is not of course cancelled for the ferromagnetic component of the spin as is the case for the antiferromagnetic component (Section 2.2)]. This is a very interesting result, and although known to occur for hydrocarbons (118) has not apparently been previously observed for 'simple' polyatomic ligands such as the carbonate ion. A more detailed analysis (117) of the data by adjusting the moment distribution on Mn, C and O (using free ion form factors for $Mn^{2+}$, $C^+$ and $O^-$, but neglecting overlap effects) to fit the measured high field form factor indicated 4% of the $Mn^{2+}$ moment to be distributed around each oxygen ion, $-4\%$ around each

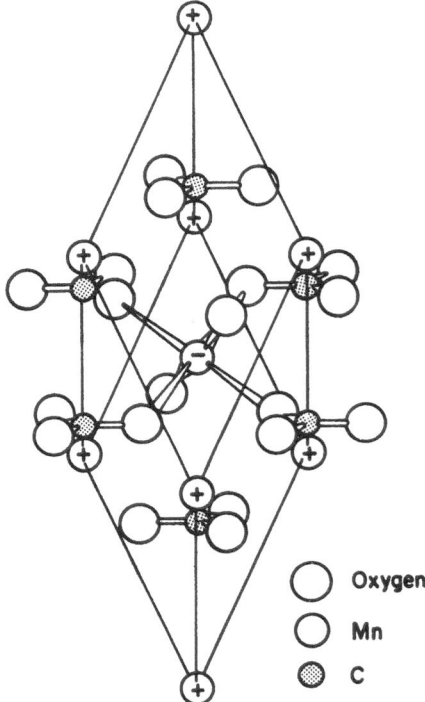

Fig. 32. The rhombohedral unit cell of $MnCO_3$. The $Mn^{2+}$ spins lie in the plane perpendicular to the trigonal axis and the spin on the atom at (000) is approximately antiparallel to the spin at $\left(\frac{1}{2} \frac{1}{2} \frac{1}{2}\right)$

⬜ Oxygen

⬜ Mn

◯ C

carbon and thus a $Mn^{2+}$ moment reduction of 8%. Also, the form factor after correction for the effects of the ligand spin showed good agreement with the free ion form factor (79). *Lindgard* and *Marshall* showed (117) that the spin transfer may be expressed in terms of $Mn^{2+} - O^{2-}$ covalency parameters:

$$1.2 \left(\frac{11}{18}f_\sigma + \frac{4}{9}f_\pi\right) = 8\% \ .$$

If $f_\sigma = f_\pi = f$, then $f \approx 6\%$. This covalency and moment reduction are larger than observed for MnO and $MnF_2$ (Table 6), which might indicate greater covalency for the carbonate [the moment reduction is similar to that observed for $\alpha$-MnS (Table 6)]. Because of the difficulties associated with the experiment and interpretation of the data however, the final result for $MnCO_3$ is probably only a qualitative estimate. The experiment is important in demonstrating the spin polarization of the carbonate ion and the power of neutron scattering to investigate such phenomena directly. The $Mn^{2+}$ moment reduction may perhaps be more accurately assessed by a measurement of the antiferromagnetic form factor.

Structure factors for spin covalently transferred to the fluoride ions in $MnF_2$ were elegantly measured using polarized neutrons by *Nathans et al.* *(57)*. $MnF_2$ is an antiferromagnet and the magnetic and chemical (rutile) unit cells are the same size (Fig. 33). The two magnetic atoms per unit cell $\left( \text{at } (000) \text{ and } \left( \frac{1}{2} \frac{1}{2} \frac{1}{2} \right) \right)$

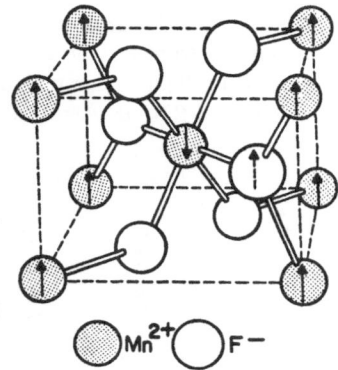

Fig. 33. The chemical and magnetic structure of $MnF_2$. The arrows indicate the direction and arrangement of the magnetic moments assigned to the manganese atoms. The broken arrow on one fluorine ion indicates the direction of the net spin transferred from the adjacent $Mn^{2+}$ ions

have opposite spins and the spin direction is along the $c$-axis (so that no (001) magnetic reflection is observed). $MnF_2$ is thus a system where both the spin configuration and orientation may readily be determined from powder scattering measurements. A simple analysis assuming spherical spin density indicates three types of reflection below $T_N$:

a) $(0, k, l)$ with $k + l$ odd will be purely magnetic,

b) all other reflections with $h + k + l$ odd will be mixed nuclear and magnetic with the nuclear scattering coming solely from the F ion nuclei,

c) reflections with $h + k + l$ even will be purely nuclear.

For a polarized beam experiment (Section 3.4) $\hat{\varkappa} \cdot \hat{\eta} = 0$ and $\boldsymbol{P}$ is parallel or anti-parallel to the spin-axis. These conditions are satisfied for $MnF_2$ for measurement of $(hk0)$ type (b) reflections with $\boldsymbol{P}$ and [001] perpendicular to $\varkappa$. For antiferromagnets however, there are further criteria to be obeyed for observation of a polarization dependence of mixed magnetic and nuclear reflections. These involve the lack of translational symmetry of the magnetic ions within the unit cell and the predominance of one type of antiferromagnetic domain within the volume of crystal exposed to the beam, and are illustrated in Fig. 34(a—c). The one-dimensional situation illustrated in Fig. 34(c), representative of MnO and NiO for example, will clearly be insensitive to the neutron polarization, but because of the lack of translational symmetry between the up and down spins, this is not the case for the domains illustrated in Figs. 34(a) and (b). In $MnF_2$

the two $Mn^{2+}$ atoms per unit cell are related by a translation of $\frac{1}{2} \frac{1}{2} \frac{1}{2}$ but the local symmetry about the manganese ions is orthorhombic and differs by a 90° rotation about the [001] axis. These properties guarantee a polarization dependence of scattering — basically the result of coherence between the $F^-$ ion nuclear scattering and the $Mn^{2+}$ magnetic scattering. The only two antiferromagnetic domains possible in $MnF_2$ are represented pictorially in one-dimension by Figs. 34(a) and (b). The observation of a polarization dependence for the type (b) reflections showed that these domains were not present equally. From a measurement of the flipping ratio of the (210) reflection for which the magnetic and nuclear structure factors are almost equal at 4.2K, it was determined that $80 \pm 5\%$ was of one domain type.

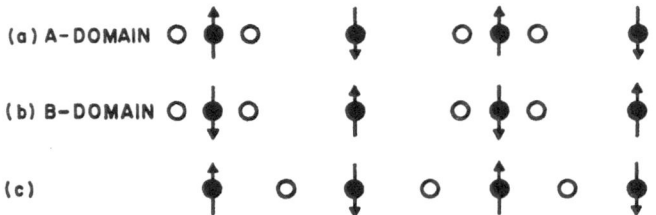

Fig. 34. One-dimensional antiferromagnetic arrays. The solid and open circles denote magnetic and non-magnetic atoms respectively. The arrows give the spin directions. Cases (a) and (b) will scatter neutrons of opposite polarization differently. Case (c) is intensitive to the polarization of the incident beam [after Ref. (57)]

A polarization dependence for type (c) (formally nonmagnetic) reflections was also observed. This 'forbidden' scattering results from the covalent transfer of spin on to the fluoride ions[28] and this spin may be examined using these reflections without interference from the very much larger spherical moment localized about the manganese ions. Ten such magnetic structure factors were measured out to $\varkappa/4\pi \approx 0.7$. These reflections do not form a complete set of data in themselves, because the 'allowed' type (b) reflections also contain a magnetic part arising from the covalent spin density. There are also contributions to the forbidden reflections from various asymmetrical spin distributions. Nevertheless, the data were interpreted by *Marshall* [unpublished but quoted in Ref. (64)] to indicate a value for $(f_\sigma + 2f_\pi + f_s)$ of 3.3%.

Such an experiment is not easy to perform and is very time consuming because of the very small magnetic structure factors for the forbidden reflections. The investigation of $MnF_2$ was important however, both in demonstrating the application of polarized neutron techniques to antiferromagnetic materials and in revealing directly the covalent spin density transferred to the fluorines [the existence of this spin density was already known of course, from NMR measure-

---

[28] The fluoride ions are coordinated by three nearest neighbor manganese ions and thus (Fig. 33) possess a net moment [Figs. 34 (a) and (b) illustrate this situation in one dimension].

ment of LHFI in $KMnF_3$ (*105*)]. It remains, in fact, the only neutron scattering investigation of covalency for $Mn^{2+} — F^-$. Because of the rather large uncertainty which is almost certainly associated with the spin transfer coefficients, an investigation to measure the moment reduction by conventional Bragg scattering for a simple manganous fluoride such as $RbMnF_3$ might be a useful contribution.

Although some measurements were made previously[29] there has, surprisingly, been little investigation of the $Fe^{3+}$ form factor until the recent polarized neutron study of YIG, of which a preliminary report has been given (*73*), although such data provide an interesting comparison with that for $Mn^{2+}$. The garnet $Y_3Fe_5O_{12}$ contains three tetrahedrally coordinated $Fe^{3+}$ ions and two octahedrally coordinated $Fe^{3+}$ ions per formula unit. These are aligned antiparallel to give a ferrimagnetic structure ($T_c = 559$ K) of considerable current technological interest for bubble domain devices. The bulk susceptibility for the unit cell could be measured but the division between octahedral and tetrahedral moments was, of course, not known, except via models. There are four classes of magnetic reflections — two have contributions from both the octahedral and tetrahedral $Fe^{3+}$ but with different sign, a third class has contributions from the tetrahedral ions only, while the fourth class is sensitive only to covalently transferred spin density on the oxygen atoms (as in $MnF_2$ this is not completely cancelled). These were all investigated with the polarized beam and the scattering from octahedral and tetrahedral $Fe^{3+}$ separated, and spin density on the oxygens observed. Values of $\langle S \rangle f(\varkappa)$ were observed at room temperature for $Fe_{oct}^{3+}$ and $Fe_{tet}^{3+}$ out to $\varkappa/4\pi = 1.0$ and the magnitudes of $\langle S \rangle$ obtained by Fourier integration (Section 3.7). The position of the yttrium ion was chosen as the background level. Integration over a spherical volume gave slightly different results to integration over a cubic volume, probably because a small quantity of the negative oxygen contribution was included. The results of the integrations varied with the dimensions of the volume of integration but the values for a cube integration with an edge of one quarter of the cell dimension gave good agreement per formula unit with the bulk magnetization data. The effective spins found thus were 3.75 and 3.70 for the octahedral and tetrahedral sites respectively.

The resulting form factors are shown in Fig. 35. For octahedral $Fe^{3+}$ the experimental curve is close to the calculated free-ion curve (*79*) but the curve for tetrahedral $Fe^{3+}$ is contracted, which might indicate $3d$ expansion, or significant spin polarization of the more diffuse $4s$ and $4p$ orbitals. The magnitudes of the moments were not discussed (*73*) and it is not clear how much of the ligand moment was included in the integration. Nevertheless the values appear to correspond roughly with those obtained by powder diffraction for other $Fe^{3+}$ oxides (*62, 71*). If it is the case that all the problems of experiment and interpretation have been successfully resolved[30] then this experiment is quite an illuminating one in providing form factors for both octahedral and tetrahedral $Fe^{3+}$.

---

[29] *e.g.*, for $Fe_3O_4$ using polarized neutrons (*119*).

[30] More recent work on YIG (*M. Bonnet*, personal communication) has included a polarized beam study at 4.2K and a profile analysis study of powder data. This gave the same values for the $Fe^{3+}$ moments as found by the Fourier integration. The data presented in Ref. (*73*) are considered essentially correct but estimation of the associated errors has involved a

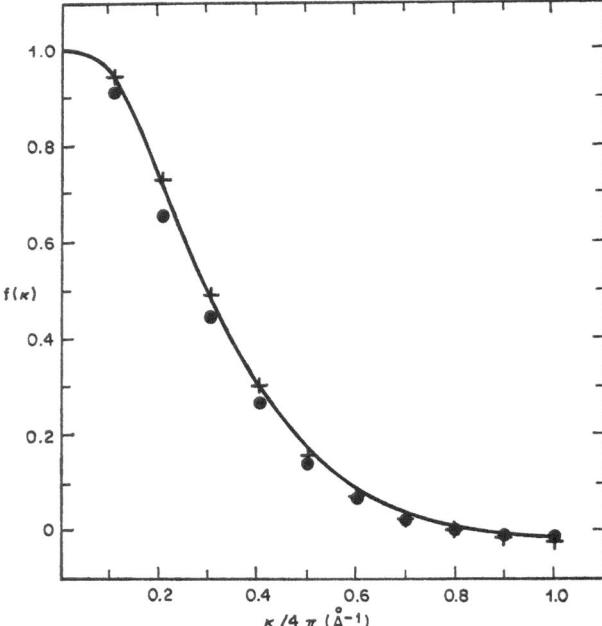

Fig. 35. The octahedral and tetrahedral form factors for $Fe^{3+}$ in $Y_3Fe_5O_{12}$ [after Ref. (73)]. The experimental points are denoted by $+$ and $\bullet$ for the octahedral and tetrahedral cases respectively. The smooth curve is the free ion form factor. The octahedral form factor follows this closely but the tetrahedral form factor is significantly contracted

It is convenient at this point to mention the one other transition metal ion form factor which has been recently investigated — that for $Co^{2+}$ in CoO. The magnetic structure of CoO has been the subject of many investigations. Although it was initially thought (120, 121), to be similar to that of NiO and MnO (Fig. 20), CoO has a small tetragonal (rather than rhombohedral) distortion below the Néel temperature and several multispin axis structures with tetragonal symmetry were also consistent (121) with the early powder data. Accurate powder diffraction data and single crystal data obtained by *van Laar* (122) narrowed the choice to the single spin axis structure or one multispin axis structure. The most recent single crystal data of *Khan* and *Erickson* (123) strongly supported the multispin axis model of *van Laar*. The latter authors determined the spherical component of the form factor and found it to be $15 - 17\%$ expanded compared to the calculated (79) 'spin only' free ion curve. The orbital moment is not quenched entirely for high-spin octahedral $Co^{2+}$ ($t_{2g}^5\ e_g^2$), however, and must be taken into account. *Mahendra* and *Khan* (124) calculated the theoretical free ion form factor

considerable effort to overcome the significant extinction effects associated with the single crystal data. An ab initio calculation performed in the light of the data obtained has also recently been presented (*E. Byrom, A. J. Freeman* and *D. E. Ellis*, Proc. Magnetism Conference, San Francisco, 1974).

for $Co^{2+}$ using the ground-state wave function for CoO determined by *Kanamori* (*125*) and found that the inclusion of the orbital moment led to a considerable expansion of the form factor [as was also the case to a lesser degree for the spin-orbit induced orbital moment in NiO (*80*) (Section 4.1)]. The agreement with the experimental form factor is now within 6% (Fig. 36). The experimental curve was determined by extrapolating the smooth curve through the values of $\langle\mu\rangle f(\varkappa)$ to give $\mu = 3.35 \pm 0.04 \ \mu_B$ at $\varkappa = 0$. and then normalizing to this value of the moment.

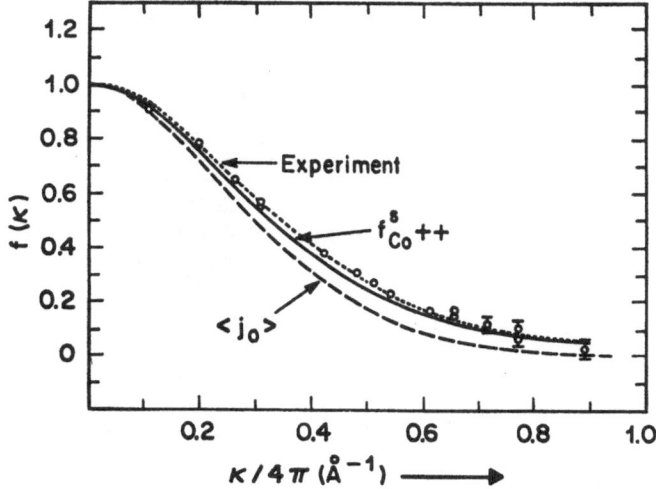

Fig. 36. The theoretically determined spherically symmetric magnetic form factor for $Co^{2+}$ in CoO. The experimental form factor and the free ion form factor $\langle j_0 \rangle$ for $Co^{2+}$ are also plotted [after Ref. (*124*)]

This work represents the current state of knowledge of the $Co^{2+}$ form factor. Neither covalency effects, the overlap contribution, nor polarization effects of inner shell or valence electrons were considered in the calculation and thus it is not possible to draw any conclusions, for example, about the expansion or contraction of the $3d$ electrons. Neither has the value of the moment been discussed with regard to covalency effects. The study of CoO is however illustrative of the great difficulties involved in determining significant information on bonding for transition metal systems with unquenched orbital moments, even though the free ion form factor computations may be performed relatively readily if the ground state wave functions are known (*1, 75, 124, 126*). The situation is less forbidding for rare earth ions, however, where the orbital moments are unquenched and the crystal field effects are small (Section 4.4).

Observed spin transfer coefficients for $Mn^{2+}$ and $Fe^{3+}$ are given in Tables 6 to 8 and individual $f_\sigma$ and $f_\pi$ for $Mn^{2+}$ and $Fe^{3+}$ in oxide and fluoride coordination, derived from the combination of LHFI and neutron data are given in Tables 7 and 8. In addition to the data given in the Tables, the neutron data for $Sr_2Fe_2O_5$

determined by profile analysis (71) indicated a covalency parameter sum of $15.3 \pm 2.1\%$ ($f_\sigma + 2f_\pi + f_s$) for octahedrally coordinated $Fe^{3+}$ and $17.5 \pm 3.3\%$ (fractional spin transfer per ligand – see Section 4.2) for tetrahedrally coordinated $Fe^{3+}$. $BiFeO_3$ has also been recently investigated (130) by both profile analysis and polarization analysis and a covalency parameter sum of $13.4 \pm 0.5\%$ determined. The four determinations for $Fe^{3+}$ (octahedral) are all fairly similar, the spread being little larger than the experimental errors, and no obvious correlations with other parameters (e.g., the nature of the other cation) stand out. For convenience,

Table 6. Spin transfer coefficients for $Mn^{2+}$ in chalgogenide lattices determined by neutron diffraction

| Host | $f_\sigma + 2f_\pi + f_s$ |
|------|---------------------------|
| MnO | $3.6 \pm 0.5\%$[a] |
| $\alpha$-MnS | $7.0 \pm 0.3\%$[b] |
| $\alpha$-MnSe | $7.5 \pm 0.3\%$[c] |
| $MnSe_2$ | $7.8 \pm 1.1\%$[c] |
| MnTe | $9.8 \pm 0.5\%$[d] |

[a] Ref. (48).
[b] Ref. (56).
[c] Ref. (61).
[d] Ref. (127).

Table 7. Spin transfer coefficients for $Mn^{2+}$ in oxide and fluoride lattices

|  | Fluorides |  | Oxides |
|---|---|---|---|
| $f_s$ | $0.52\%$ (in $KMnF_3$, NMR)[a] $0.52\%$ (in $RbMnF_3$, NMR)[b] $0.55\%$ (in $KMgF_3$, ESR)[c] | $f_s$ | $0.8 \pm 0.002\%$ (in MgO, ENDOR)[e] |
| $f_\sigma - f_\pi$ | $0.2\%$ (in $KMnF_3$, NMR)[a] $0.3\%$ (in $RbMnF_3$, NMR)[b] $0.3\%$ (in $KMgF_3$, ESR)[c] | $f_\sigma - f_\pi$ | $0.8 \pm 0.6\%$ (in MgO, ENDOR)[e] |
| $f_\sigma + 2f_\pi + f_s$ | $3.3\%$ (in $MnF_2$, neutrons)[d] | $f_\sigma + 2f_\pi + f_s$ | $3.6 \pm 0.5\%$ (in MnO, neutrons)[f] |
| $f_s = 0.5\%$ |  | $f_s = 0.8 \pm 0.002\%$ |  |
| $f_\sigma = 1.1\%$ (probable error $0.5\%$) |  | $f_\sigma = 1.5 \pm 0.6\%$ |  |
| $f_\pi = 0.8\%$ (probable error $0.5\%$) |  | $f_\pi = 0.7 \pm 0.4\%$ |  |

[a] Ref. (105).
[b] Ref. (128).
[c] Ref. (84).
[d] Ref. (57).
[e] Ref. (102).
[f] Ref. (48).

Table 8. Spin transfer coefficients for $Fe^{3+}$ in oxide and fluoride lattices

|  | | Fluorides | | Oxides |
|---|---|---|---|---|
| $f_s$ | | 0.80% (in $KMgF_3$, ESR)[a] <br> 0.80% (in $K_2NaGaF_6$, ESR)[b] | $f_s$ | 1.05 ±0.002% (in MgO, ENDOR)[d] |
| $f_\sigma - f_\pi$ | | 3.4% (in $KMgF_3$, ESR)[a] <br> 3.4% (in $K_2NaGaF_6$, ESR)[b] | $f_\sigma - f_\pi$ | 5.7 ±0.7% (in MgO, ENDOR)[d] |
| $f_\sigma + 2f_\pi + f_s$ | | 6.2 ±0.7% (in $FeF_3$, neutrons)[c] | $f_\sigma + 2f_\pi + f_s$ | 11.8 ±0.5% (in $LaFeO_3$, neutrons)[e] |
| | | | | 12.9 ±1.0% (in $YFeO_3$, neutrons)[e] |
| | $f_s = 0.8\%$ | | $f_s = 1.05\%$ | |
| | $f_\sigma = 4.1 \pm 0.6\%$ | | $f_\sigma = 7.6 \pm 0.8\%$ | |
| | $f_\pi = 0.7 \pm 0.3\%$ | | $f_\pi = 1.9 \pm 0.4\%$ | |

[a] Ref. (84).
[b] Ref. (129).
[c] Ref. (101).
[d] Ref. (102).
[e] Ref. (62)[f].
[f] These values have been corrected to the revised scattering length of germanium [Ref. (49)].

we have continued to use the data of Ref. (62) to determine individual $f_\sigma$ and $f_\pi$ by comparison with resonance data (Table 8). The data for tetrahedral $Fe^{3+}$ in $Sr_2Fe_2O_5$ follow the trend found for tetrahedral $Fe^{3+}$ in YIG and tetrahedral $Co^{2+}$ in $Co_3O_4$ and $CoRh_2O_4$ (Section 4.2), i.e., similar moment reductions to octahedrally coordinated ions indicating greater covalency per 'bond' in the tetrahedral cases. The investigation (72) of $MnRh_2O_4$ ($T_N \approx 20$ K) at 1.2 K seems to indicate a somewhat larger moment reduction (17.5 ±3%) for the tetrahedral $Mn^{2+}$ ion than found for octahedral $Mn^{2+}$ in MnO. Although this preliminary value may be revised after data analysis is fully completed, it is clear that the investigation of tetrahedrally coordinated ions has started to provide very interesting information.

$Mn^{2+}$ LHFI was investigated (84) in alkali halides of increasing lattice parameter and $f_\sigma - f_\pi$ was found to increase from 0.7% for LiF (M—F distance = 2.009 Å) to 2.1% for KF (M—F distance = 2.673 Å). This effect is in keeping with the directional properties of the $3d_\sigma$ and $3d_\pi$ orbitals. The isoelectronic $d^5$ ion $Cr^+$ had a negative value of $f_\sigma - f_\pi$ both in NaF (−0.6%) and in $KMgF_3$ (−1.5%) (84).

The data of Table 6[31]), determined by internal calibration on polycrystalline materials show the expected increase in covalency from oxide to sulphide (~2x) and from sulphide to telluride. The sulphide, selenide and diselenide are all seen

---

[31]) $Mn^{2+}$, with a half-filled $3d$ shell, has several chemical properties which indicate relatively low covalency in many situations compared to other divalent transition metal ions, reflecting the 'stability of the half-filled shell'. The localized electron behavior of the chalcogenides is one such property — the chalcogenides of the other ions generally show more complicated properties (with metallic behavior or metal-semiconductor transitions).

to have similar covalency. These values are in rather good accord with the ESR data for $Co^{2+}$ halides (Section 4.1). The similar behavior, in both examples, of third and fourth row anions is consistent with many chemical properties and the similarity of the covalency for the monatomic and diatomic selenide ions is not unexpected.

From Table 7 we see that there is no significant difference in covalency parameters for $Mn^{2+} - F^-$ and $Mn^{2+} - O^{2-}$. This agrees with the neutron diffraction data for $Ni^{2+}$ (Section 4.1). Although $f_\sigma$ is much less than for $Ni^{2+}$, the total ligand-to-metal charge transfer by $\sigma$ and $\pi$ bonding from Eq. (2.23) is very similar for the two ions ($\sim 0.2e$).

For $Fe^{3+}$ (Table 8) this similarity for fluoride and oxide coordination no longer holds, as might be expected in a more covalent, higher oxidation state situation. Such an effect was observed also for $Cr^{3+}$ (Table 4, Section 4.2), but in contrast to $Cr^{3+}$ we see that the covalent interaction for $Fe^{3+}$ is a predominantly $\sigma$-bonding one. Apparently the relative importance of ligand-to-metal $\pi$ bonding decreases sharply across the first transition series, especially for divalent or higher oxidation state ions. As expected, both $Fe^{3+}$ fluorides and oxides are considerably more covalent than the $Mn^{2+}$ salts. Charge transfers of $\sim 0.4e$ and $\sim 0.8e$ respectively are estimated from Eq. (2.23).

The data of Tables 6—8 provide a consistent picture. The low value of $f_\sigma$ for $Mn^{2+}$ does not seem to be anomalous in view of the additional $\pi$-bonding pathway which is not present for $Ni^{2+}$. No account of spin polarization has been taken in this analysis however. As already mentioned 4s spin polarization was investigated theoretically by *Hubbard et al.* (36). The effect may be expected to be most significant for the $d^5$ high-spin situation but the experimental data, in contrast to the situation for $d^3$ ions, do not permit any direct estimate to be made. Spin polarization will decrease the value of $f_\sigma$ observed by LHFI (so decreasing $f_\sigma - f_\pi$) and decrease the moment reduction observed by neutrons. The spin-unrestricted calculations which have been made (36, 76, 93, 94) do not agree on the magnitude of the effect, but seem to agree that it is quite small and not of sufficient magnitude to qualitatively alter the conclusions discussed. Because of the increased $3d - 4s$ splitting spin polarization is agreed to be of smaller magnitude for $Fe^{3+}$ than $Mn^{2+}$. On the other hand for $Cr^+$, a relatively large effect may help to explain the negative value of $f_\sigma - f_\pi$ observed by spin resonance (84).

*Soules* and *Richardson* calculated (76) $f_\sigma = 1.5\%$ and $f_\pi = 0.7\%$ for $Mn^{2+} - F^-$, in quite good agreement with the values of Table 7. *Brown* und *Burton* (93) found $\sigma$-bonding dominant for both $Fe^{3+}$ and $Mn^{2+}$ so that their values of $f_\sigma - f_\pi$ (calculated for $Fe^{3+}$ in $K_2NaFeF_6$ and for $Mn^{2+}$ in $KMnF_3$) of 5.14% and 1.98% were reasonable for $Fe^{3+}$ but too high for $Mn^{2+}$. Spin density coefficients were not calculated for the $FeO_6^{9-}$ cluster (37) (Fig. 4) but in a multiple scattering $X\alpha$ calculation for fluoride coordination, *Larsson* and *Connolly* (94) found similar values of $f_\sigma - f_\pi$ to *Brown* and *Burton*. The multiple scattering calculation seems to indicate greater spin polarization of the bonding $e_g$ and $t_{2g}$ orbitals than the other calculations. The calculations of Refs. (93) and (94) for $Cr^{3+}$ have already been discussed (Section 4.2).

Thus, although not yet providing a well-rounded picture of covalency for $Mn^{2+}$, the calculations so far attempted do not appear to conflict seriously with the interpretation of Tables 7 and 8, and do support the interpretation of dominant $\sigma$ covalency determined for $Fe^{3+}$. Other experimental evidence has been interpreted, however, as indicating greater covalency for $Mn^{2+} - O^{2-}$ than for $Mn^{2+} - F^-$, and a value of $f_\sigma$ for $Mn^{2+} - F^-$ similar to that found for $Ni^{2+} - F^-$.

The hyperfine splitting ($A$) observed by electron spin resonance for $Mn^{2+}$ doped into oxides is $\sim 10\%$ less than for $Mn^{2+}$ in fluoride hosts [for a Table of values see Ref. (35)]. Also values of $A$ for $Cr^+$, $Mn^{2+}$ and $Fe^{3+}$, when plotted against $C/n$ lie on a straight line or a smooth curve, indicating an influence of 'covalency' on the magnitude of $A$. $C$ is the 'covalency parameter' determined from the Pauling atomic electronegativities of the dopant ion and the ligand using the relation of *Hannay* and *Smyth* (131) (even though these authors were concerned with rationalizing the dipole moments of diatomic molecules), and n is the coordination number of the cation. Such a relation was first derived by *Matumura* (132) and the curves are shown in more detail in Ref. (35). For these $d^5$ ions the hyperfine field at the nucleus is given by the contact field ($H_s$) due to spin polarization of s electrons. The value of $H_s$ extrapolated for zero covalency for $Mn^{2+}$ agrees well with the value calculated for the free ion (133) ($- 700$ kOe). It was originally thought that the reduction of $A$ with increasing 'covalency' was the result of increasing spin transfer on to the ligands, so reducing the spin polarization of inner shell metal s orbitals. But in this case $A$ might have been expected to show a dependence on the host cation-to-ligand distance as does $f_\sigma - f_\pi$ for $Mn^{2+}$ (84), but it does not [e.g., in MgO, CaO and SrO at 4.2K, $A$ ($\times 10^4 cm^{-1}$) is $81.5 \pm 0.2$, $81.6 \pm 0.001$ and $80.9 \pm 0.2$ respectively (134)]. More recently *Simanek* and *Müller* have suggested (35) that the dominant effect leading to the reduction of $A$ is associated with the radial polarization of the covalently occupied up- and down-spin 4s electrons. Thus, the nuclear hyperfine interaction seems to have quite a different origin to the spin densities determined by LHFI and neutron diffraction, and the two sets of data are not necessarily inconsistent.

The hyperfine fields at the $^{111m}Cd$ nucleus doped into antiferromagnetic $RbMnF_3$, $KCoF_3$ and $KNiF_3$ were recently found (135) to be 113.5 KOe, 74.4 KOe and 105.6 KOe respectively. The values were analysed as arising from supertransferred spin densities in the Cd orbitals, which, because of the symmetry of the perovskite lattice were assumed to be proportional to the spin density in the nearest neighbor fluoride $p_\sigma$ orbitals ($f_\sigma$). Thus these hyperfine fields should be proportional to $f_\sigma$ and the data appeared to show, therefore, that $f_\sigma$ for $Mn^{2+} - F^-$ was indeed as great as for $Ni^{2+} - F^{-32}$). If this were the case, however, $f_\sigma + 2f_\pi + f_s$ would be $\sim 12\%$ — as large as observed for $Fe^{3-} - O^{2-}$ by neutron diffraction, and quite inconsistent with the neutron diffraction data for MnO (48). *Rinneberg* and *Shirley* thus suggested (135) that some unknown systematic error caused the covalency parameters determined by neutron diffraction to be anomalously low. This does not seem very likely and it was not suggested where the extra scattering for MnO, for example, was supposed to come from if the measured co-

---

[32]) It was not thought worthy of comment that $f_\sigma$ for $Co^{2+}$—$F^-$ [given as 2.6%, in good agreement with the estimate (87) of *Thornley et al.* (2.4 $\pm$ 1%)] was therefore presumably 'anomously' low in turn.

valency parameters were far too low. Errors in the zero-point spin deviation could not give rise to such large discrepancies.

It does appear, however, that effects involving the $4s$ orbitals may have been neglected in interpreting the hyperfine field data. Polarization of the metal core $s$ orbitals by overlap with ligand $\sigma$ orbitals was indeed the dominant mechanism calculated by *Huang et al.* (*136*) for $Mn^{2+}$ hyperfine fields in $KMnF_3$ and $MnO$, but a different mechanism involving the direct transfer from the valence orbitals of a neighboring metal to the empty outer s orbitals has been shown (*137*) to be probably dominant for the supertransferred hyperfine interactions (STHFI) measured by the Mössbauer effect at $^{119}Sn^{4+}$ and $^{121}Sb^{5+}$ doped into YIG and nickel ferrite respectively. In addition, *Lau* and *Newman* have recently shown (*138*), on the basis of the APW band calculations of *Mattheiss* (*97*) that direct $4s - 3d$ interaction appears to be dominant in producing the STHFI in $MnO$. There is, incidentally, a rather poor mismatch of ionic radii between $Cd^{2+}$ (1.09 Å in fluorides) and $Mn^{2+}$, $Co^{2+}$ and $Ni^{2+}$ (0.960Å, 0.875Å and 0.840Å in fluorides) (*139*).

Clearly the details of bonding have not yet been completely delineated for $Mn^{2+}$, although a consistent picture is obtained from the consideration of the neutron and LHFI data. More computational effort is required to explain in detail the origins of the various covalency-related effects discussed. As is the case for the ions discussed in the previous sections, more neutron diffraction data for $Fe^{3+}$ and $Mn^{2+}$ (both moment reduction and form factor) for ligands other than oxide and fluoride would be desirable.

$Mn^{2+}$ has been discussed in this section and $Mn^{4+}$ in Section 4.2. It is pertinent at this stage to mention investigations of $Mn^{3+}$ ($d^4$ high-spin). $LaMnO_3$ was studied for covalency by *Nathans et al.* (*64*), but no covalent moment reduction was found. The $MnO_6^{9-}$ octahedra are considerably distorted in $LaMnO_3$ as a result of the Jahn-Teller distortion associated with the single $e_g$ $\sigma$-antibonding electron. This also leads to a different type of magnetic ordering (*140*) (A-type, Fig. 22) with ferromagnetic sheets in the (001) planes and antiferromagnetic coupling between the sheets along the $c$-axis[33]). A consequence of this magnetic structure is that the ligand moment is not completely quenched and a consequence of the spin orientation ($\perp[001]$) is that an (001) magnetic reflection is observed at $\varkappa/4\pi = 0.065$. Comparison of the (001) intensity with the intensity of the next magnetic reflection [(111) at $\varkappa/4\pi = 0.14$] might allow the observation of an effect on the (001) due to the presence of the ligand moment. It is not clear why $LaMnO_3$ should not otherwise show a moment reduction and this compound and $MnF_3$ which has similar crystallographic and magnetic ordering (*142*) are currently being reinvestigated by powder diffraction (*143*).

## 4.4 Rare Earth Ions

Apart from the data already discussed for salts of the $3d$ transition metal ions, significant bonding information has otherwise been obtained only for a few rare earth ions. The magnetic and crystallographic properties of many $4d$ and $5d$

---

[33]) The relationship between the crystal symmetry and the magnetic properties has been discussed by *Goodenough et al.* (*141*).

transition metal compounds (metal cluster formation, metallic behavior, metal-metal bonding, etc.), the invariably low spin behavior associated with the strong crystal fields[34]), and the sharply dropping form factors [for any value of $\varkappa/4\pi$, $f(\varkappa)$ will be smaller for a $4d$ or $5d$ ion than for the corresponding $3d$ ion because of the greater spatial extent of the magnetic wavefunctions] together make many of these compounds unattractive, or anyway very difficult to study by magnetic neutron scattering. Only $MoF_3$, $(4d^3)$ which has a similar magnetic structure to $CrF_3$ and $FeF_3$ *(144)* stands out as an attractive candidate for study by conventional techniques.

The magnetic ordering of a number of actinide compounds has been investigated, but the electron configuration itself is often difficult enough to determine, as the outer $5f$, $6d$ and $7s$ shells are of similar energy, and while the $5f$ electrons may be localized, $6d$ and $7s$ electrons will probably be collective.

The larger crystal fields observed for actinide ions compared to rare earth ions and the comparative lack of accurate theoretical calculations leads to further problems and greater characterization of the electronic states will be necessary before effects due to bonding can be readily distinguished. The most recent and comprehensive study of an actinide compound is that of US by *Wedgwood* *(19)*. US is ferromagnetic and the form factor was determined to $\varkappa/4\pi \sim 0.95$ using polarized neutrons and was of nearly spherical symmetry. The results did show that there was very little moment on the sulphur atoms ($< 0.02\ \mu_B$ per atom) so that covalent bonding of the type discussed in this article is apparently small. Comparison of the moment determined by extrapolation of the form factor to $\varkappa = 0$ (1.7 $\mu_B$ per atom) with the bulk magnetization data (which gave 1.54 $\mu_B$ per atom) indicated a negative conduction polarization in the $6d$ or $7s$ bands[35]). The best fit to the form factor was given by a model with $5f^2$ configuration and an exchange field of comparable magnitude to the crystal field. The observation of ligand spin densities (or the lack of them) as was done for US is probably the most direct way of investigating bonding effects in such complicated materials as $4d$ and $5d$ transition metal and actinide compounds.

The chemical behavior of the trivalent rare earths, the low magnetic ordering temperatures of most rare earth compounds with unfilled $4f$ shells[36]), and the ligand hyperfine interactions observed in spin resonance measurements[37]) all indicate predominantly ionic behavior. This is presumably the result of the shielding of the $4f$ electrons from the chemical environment by the $5s^2 5p^6$ shell. This shielding is also reflected in the narrow-line optical spectra of the trivalent

---

[34] Low spin systems are, of course, amenable to study by NQR (Section 2.2) and several $4d$ and $5d$ chlorides, bromides and iodides have been investigated [see Ref. *(4)*].

[35] Similar effects observed for Fe, hexagonal Co, Ni and Gd were mentioned in Section 3.7.

[36] Generally, $< 4K$. In a few situations (*e.g.*, EuO—EuTe) collective electron interactions involving outer $d$ or $s$ orbitals are also present inducing significantly higher magnetic ordering temperatures.

[37] LHFI data for rare earths are much less extensive than for $3d$ salts. Some data on the isotropic interaction are summarized in Ref. *(145)*. Except for $Tm^{2+}$ in $CaF_2$ studied in that work, the direct $4f$ interaction, which leaves parallel unpaired spin on the ligands, is weaker than the interactions between the ligands and the spin polarized $5s^2 5p^6$ shell which transfer antiparallel spin to the ligand.

rare earths and the weak crystal field splittings observed. Even though it appears that an electrostatic crystal field model is no more appropriate for describing rare earth crystal field splittings than for d-series ions (146), recent neutron diffraction data for $Gd_2O_3$ (21) and $Tb(OH)_3$ (147, 53) seem to confirm that with regard to $4f$ bonding at least, these trivalent rare earth compounds are indeed significantly less covalent than the $3d$ transition metal compounds discussed above.

The form factor for polycrystalline $Gd_2O_3$ was determined (21) at room temperature to $\varkappa/4\pi \sim 0.5$ by polarization analysis. The form factor for single crystal $Tb(OH)_3$ was determined (147) at 2.6K (below the ferromagnetic ordering temperature of 3.72K) and in the paramagnetic phase at 90K using the polarized beam, with an applied field of 12.0 $\pm 0.5$ kOe at both temperatures. A search was made for spin density transferred to the oxygen atoms which could be measured by observing a polarization dependence for formally non-magnetic reflections [as was done for YIG and $MnF_2$ (Section 4.3)]. In no case was there any significant deviation of the flipping ratio from unity, indicating that any moment located at the oxygen site must be less than 0.01 $\mu_B$.

The localized spherical $4f$ form factor determined for metallic gadolinium (21) demonstrated rather elegantly that the radial distribution of the $4f$ electrons was significantly expanded relative to the non-relativistic Hartree-Fock wave functions (148, 149) which had been used previously in discussing the magnetic and electrical properties of rare earth ions. The fully relativistic Dirac-Fock calculations of *Freeman* and *Desclaux* (150) were however in very good agreement with the experimental data. The relativistic contraction of the core electrons increases slightly the shielding of the $4f$ electrons from the nucleus causing the $4f$ radial density to expand relative to the non-relativistic Hartree-Fock result. A very similar result was obtained for $Tb(OH)_3$, where the relativistic form factor for $Tb^{3+}$ (151) fell directly on to the best fit to the experimental points (Fig. 37). These experiments demonstrate once again the particularly stringent test given by magnetic form factor data for theoretical wavefunctions.

The $Gd_2O_3$ form factor appeared, on the other hand, to follow the non-relativistic curve (149) for $Gd^{3+}$ and this apparent expansion of the $4f$ density was presumed (150) to result from the effect of covalent bonding. However it has since been found (152) that there is still a small amount of residual short range magnetic order at 300K, even though the magnetic ordering temperature is below 4.2K. The small corrections for this effect and for a very small revision of the absolute scale factor resulted in a revised form factor mid-way between the relativistic and non-relativistic calculations, with error bars including both calculations.

Although variable valency is not as pronounced with the rare earth metals as with the transition metals, higher oxidation states of some rare earths may be prepared. Tetravalent $Ce^{4+}$, $Pr^{4+}$ and $Tb^{4+}$ are particularly easy to stabilize in the perovskites $BaMO_3$, and polycrystalline $BaPrO_3$ and $BaTbO_3$ were studied (28) by magnetic susceptibility and neutron diffraction. No magnetic ordering was observed in $BaPrO_3$ down to 2K[38]), but $BaTbO_3$ was found to order anti-

---

[38]) Although more recent [141]Pr Mössbauer measurements indicate a magnetic transition at 11.0 $\pm 0.4$K (153).

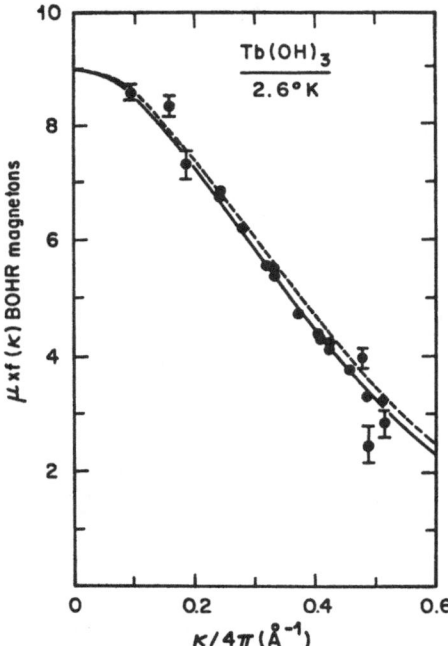

Fig. 37. Experimental $\mu f(\varkappa)$ values for Tb(OH)$_3$ in the ordered state at 2.6 K, after making extinction corrections. The broken curve is derived from nonrelativistic free ion wave functions. The solid curve is the best fit to the experimental points and falls directly on the relativistic free ion form factor. All curves are normalized to a magnetic moment of 8.9 $\mu_B$/Tb atom [after Ref. (151)]

ferromagnetically with G-type ordering (Fig. 22) at 37K (Fig. 8). Greater co-valency is anticipated for the higher oxidation state (Tb$^{4+}$, $f^7$, is isoelectronic with Gd$^{3+}$) and this high ordering temperature might reflect such an effect. The crystal field splittings are considerably larger for tetravalent than for trivalent rare earth ions, and ENDOR of Gd$^{3+}$ and Tb$^{4+}$ in ThO$_2$ (154) indicated an increase in covalency in going to the higher oxidation state. The $\sim 180°$ superexchange link is also a favorable feature for strong magnetic coupling in BaTbO$_3$, however, and indeed, TbO$_2$, where the metal-oxygen-metal link is tetrahedral has an or-dering temperature of only 3K.

Because of the high spin and spherical symmetry of the $f^7$ orbital singlet ground state and the relatively simple magnetic and nuclear structures of BaTbO$_3$ it was possible to determine $\langle S \rangle f(\varkappa)$ from the polycrystalline data out to $\varkappa/4\pi = 0.45$ at 4.2K (Fig. 38). The scale factor was determined from a nuclear structure factor refinement at 4.2K. Although no Tb$^{4+}$ form factor had been published, a value of 0.950 was estimated for the lowest angle magnetic reflection by extrapolation from the non-relativistic Hartree-Fock calculations for Eu$^{2+}$ and Gd$^{3+}$ (149). The value of $\langle S \rangle$ determined was 3.33 $\pm 0.02$, a reduction of 5.0 $\pm 0.6\%$ from the free ion value. Estimating an approximate zero-point spin reduction of 2% left a net reduction of $\sim 3\%$ which possibly reflected the presence of covalent

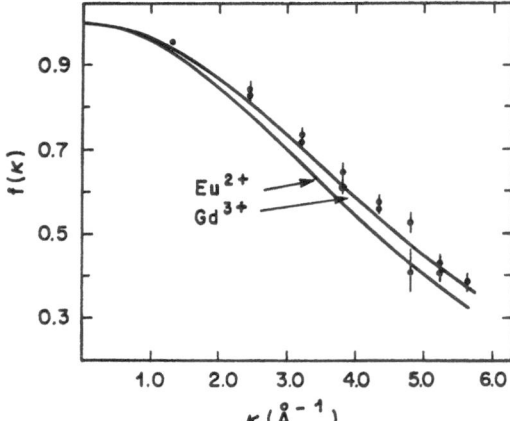

Fig. 38. The magnetic form factor for $Tb^{4+}$ in $BaTbO_3$. The experimental points have been normalized to the value of the $Tb^{4+}$ form factor at the lowest angle reflection ($\varkappa = 1.27$) determined by extrapolation from the nonrelativistic $Eu^{2+}$ and $Gd^{3+}$ theoretical free ion curves (shown as full curves) [after Ref. (28)]

interactions. Use of a relativistic form factor would decrease the moment reduction by $\sim 1\%$. Clearly, however, any moment reduction in $BaTbO_3$ is less than observed even for divalent $Mn^{2+}$ in MnO, so that even for a tetravalent rare earth ion there is as yet no evidence of significant covalency effects affecting the $4f$ charge and spin distribution.

Of course the bonding situation for $Tb^{4+}$ in $BaTbO_3$ is possibly more complicated than for $Mn^{2+}$ in MnO because the $4f$ electrons are almost certainly more tightly bound than the oxygen $2p$ electrons and spin polarization and covalency effects involving the outer $s$, $p$ and $d$ orbitals may also play a role (155). Calculations for $Tb^{4+}$ and other rare earth ions are in progress (156).

## 5. Summary

The investigations leading up to, and the progress that has been made in the ten years since the publication of the paper of *Hubbard* and *Marshall* (26), have been reviewed. These experiments have provided much new and interesting information on the bonding of magnetic ions in crystals, particularly for combination with oxide and fluoride ions. Form factor measurements have provided quantitative information on the distribution of unpaired electrons in $3d$ transition metal, and rare earth compounds, and moment reductions, in combination with ligand hyperfine interaction data, have allowed the estimation of $\sigma$ and $\pi$ covalency parameters within the framework of the molecular orbital model. The existence of spin polarization effects has been demonstrated for $d^3$ ions. In several instances covalently transferred spin on ligand atoms has been directly revealed in polarized neutron studies using single crystals.

The agreement with calculation is sometimes satisfactory but the correct interpretation of the data in other cases remains in doubt. The importance, or existence, of expanded $3d$ orbitals, the magnitude of spin polarization effects, the comparative behavior of ions in concentrated and dilute systems, and the behavior of $Mn^{2+}$ are cases in point.

Also, a significant amount of bonding information has been collected so far only for orbital singlet ground state octahedral (and occasionally tetrahedral) $3d$ ions and one or two rare earth ions. However, more extensive use of the newer techniques such as profile analysis with polycrystalline samples and polarization analysis on paramagnetic and magnetically ordered materials as well as continued measurement of form factors by conventional polarized beam techniques will undoubtedly extend the study to other metal ions and to many other ligands. Such data will provide a more comprehensive test for theoretical models of bonding than is perhaps possible with any other experimental technique. It is hoped that this review might stimulate chemists to become involved in this exciting and productive field of investigation.

*Acknowledgements.* I thank *B. T. M. Willis* for a critical reading of the manuscript, and acknowledge productive discussions with *F. A. Wedgwood* and *C. Infante*, and helpful comments from *H. Fuess* and *M. T. Hutchings*.

# 6. References

1. *Marshall, W., Lovesey, S. W.:* Theory of thermal neutron scattering. Oxford: Clarendon Press 1971.
2. Chemical applications of thermal neutron scattering (Ed. *B. T. M. Willis*). London: Oxford University Press 1973.
3. *Cox, D. E.:* Table of antiferromagnetic materials studied by neutron diffraction. Brookhaven National Laboratory, Upton, N.Y. 11973, BNL Report 13822, June 1969, *E. F. Bertaut,* Ann. Phys. (Paris) *7,* 203 (1972).
4. *Tofield, B. C.:* The study of electron distributions in inorganic solids — a survey of techniques and results. Progr. Inorg. Chem., *20* to be published.
5. *Shull, C. G., Yamada, Y.:* J. Phys. Soc. Japan *17,* Suppl. BIII, 1 (1962).
6. *Mook, H. A.:* Phys. Rev. *148,* 495 (1966).
7. *Moon, R. M.:* Phys. Rev. *136,* A195 (1964).
8. *Nathans, R., Paoletti, A.:* Phys. Rev. Letters *2,* 254 (1959).
9. *Goodenough, J. B.:* Progr. Solid State Chem. *5,* 145 (1971).
10. *Anderson, P. W.:* Magnetism, Vol. I, p. 25 (eds. *G. T. Rado* and *H. Suhl*). New York: Academic Press 1963.
11. *Owen, J., Thornley, J. H. M.:* Rept. Progr. Phys. *29,* 675 (1966).
12. *Goodenough, J. B.:* Magnetism and the chemical bond. Interscience 1963.
13. *Goodenough, J. B., Longo, J. M., Kafalas, J. A.:* Mat. Res. Bull. *3,* 471 (1968).
14. *Moon, R. M.:* Phys. Rev. Letters. *25,* 527 (1970).
15. *Hastings, J. M., Corliss, L. M.:* IBM J. Res. Develop. *14,* 227 (1970).
16. *Gautier, F., Krill, G., Lapierre, M. F., Robert, C.:* J. Phys. C: Solid State Phys. *6,* L320 (1973).
16a. *Coey, J. M. D., Brusetti, R., Kallel, A., Schweizer, J.* and *Fuess, H.:* Phys. Rev. Letters, *32,* 1257 (1974).
17. *Tofield, B. C.:* D. Phil. Thesis, University of Oxford 1969.
18. *Takeda, T., Yamaguchi, Y., Watanabe, H.:* J. Phys. Soc. Japan *33,* 967 (1972).
19. *Wedgwood, F. A.:* J. Phys. C: Solid State Phys. *5,* 2427 (1972).
20. *Tofield, B. C., Fender, B. E. F.:* J. Phys. C: Solid State Phys. *4,* 1279 (1971).
21. *Moon, R. M., Koehler, W. C., Cable, J. M., Child, H. R.:* Phys. Rev. B *5,* 997 (1972).
22. *Wedgwood, F. A.:* Proc. Roy. Soc. (London), to be published.
23. *Jørgensen, C. K.:* Progr. Inorg. Chem. *4,* 73 (1962).
24. *Owen, J., Stevens, K. W. H.:* Nature *171,* 836 (1953).
25. *Thornley, J. H. M.:* J. Phys. C: Solid State Phys. *1,* 1024 (1968).
26. *Hubbard, J., Marshall, W.:* Proc. Phys. Soc. (London) *86,* 561 (1965).
27. *Alperin, H. A.:* J. Phys. Soc. Japan, Supl. BIII *17,* 12 (1962).
28. *Tofield, B. C., Jacobson, A. J., Fender, B. E. F.:* J. Phys. C: Solid State Phys. *5,* 2887 (1972).
29. *Jacobson, A. J.:* in Ref. 2, p. 270.
30. *Ballhausen, C. J.:* Introduction to ligand field theory. New York: McGraw-Hill 1962.
31. *Watson, R. E., Freeman, A. J.:* Phys. Rev. *134,* A1526 (1964).
32. *Simanek, E., Sroubek, Z.:* In: Electron paramagnetic resonance, p. 535 (ed. *S. Geschwind*). New York: Plenum 1972.
33. *Mattheiss, L. F.:* Phys. Rev. B *5,* 306 (1972).
34. *Phillips, J. C.:* Bonds and bands in semiconductors (New York—London: Academic Press, 1973), Rev. Mod. Phys., *42,* 317 (1970).
35. *Simanek, E., Müller, K. A.:* J. Phys. Chem. Solids *31,* 1027 (1970).
36. *Hubbard, J., Rimmer, D. E., Hopgood, F. R. A.:* Proc. Phys. Soc. *88,* 13 (1966).
37. *Tossell, J. A., Vaughan, D. J., Johnson, K. H.:* Nature (Phys. Sci.) *244,* 42 (1973).

38. *Evans, S., Hamnett, A., Orchard, A. F., Lloyd, D. R.:* Discussions Faraday Div. Chem. Soc. *54*, 219 (1972).
39. *Jørgensen, C. K.:* Chimia *27*, 203 (1973); Struct. Bonding *13*, 199 (1973).
40. *Owen, J., Thornley, J. H. M., Windson, C. G.:* Proc. Phys. Soc. (London) *85*, 103 (1965).
41. *Hirakawa, K., Ikeda, H.:* J. Phys. Soc. Japan *35*, 1608 (1973), Phys. Rev. Letters, *33*, 374 (1974).
42. *Bersohn, R., Shulman, R. G.:* J. Chem. Phys. *45*, 2298 (1966).
43. *Townes, C. H., Dailey, B. P.:* J. Chem. Phys. *17*, 782 (1949).
44. *Rinneberg, H., Hartmann, H.:* J. Chem. Phys. *52*, 5814 (1970).
45. *Tsay, F. D., Helmholz, L.:* J. Chem. Phys. *50*, 2642 (1969).
46. *Carlson, E. H.:* Phys. Letters A *29*, 696 (1969).
47. *Asker, W. J., Scaife, D. E., Watts, J. A.:* Australian J. Chem. *25*, 2301 (1972).
47a. The use of inelastic neutron scattering techniques to probe the electronic states of inorganic materials has been described by *M. T. Hutchings*, Proc. Nato Advanced Study Institute on Electronic States of Inorganic Compounds: New Experimental Techniques, Oxford, 8—18 September, 1974 (Ed. *P. Day*), to be published.
48. *Jacobson, A. J., Tofield, B. C., Fender, B. E. F.:* J. Phys. C: Solid State Phys. *6*, 1615 (1973).
49. *Bacon, G. E.:* Acta Cryst. A *28*, 357 (1972).
50. *Hewat, A. W.:* J. Phys. C: Solid State Phys. *5*, 1309 (1972).
51. *Arndt, U. W., Willis, B. T. M.:* Single crystal diffractometry. Cambridge: Cambridge University 1966.
52. *Zachariasen, W. H.:* Acta Cryst. *23*, 558 (1967).
53. *Lander, G. H., Brun, T. O.:* Acta Cryst. *29*, 684 (1973).
54. *Fender, B. E. F.:* in Ref. 2. p. 250.
55. *Shirane, G.:* Acta Cryst. *12*, 282 (1959).
56. *Fender, B. E. F., Jacobson, A. J., Wedgwood, F. A.:* J. Chem. Phys. *48*, 990 (1968).
57. *Nathans, R., Alperin, H. A., Pickart, S. J., Brown, P. J.:* J. Appl. Phys. *34*, 1182 (1963).
58. *Brown, P. J., Forsyth, J. B.:* Proc. Phys. Soc. (London) *92*, 125 (1967).
59. *Moon, R. M., Riste, T., Koehler, W. C.:* Phys. Rev. *181*, 920 (1969).
60. *Moon, R. M., Riste, T., Koehler, W. C., Abrahams, S. C.:* J. Appl. Phys. *40*, 1445 (1969).
61. *Jacobson, A. J., Fender, B. E. F.:* J. Chem. Phys. *52*, 4563 (1970).
62. *Tofield, B. C., Fender, B. E. F.:* J. Phys. Chem. Solids *31*, 2741 (1970).
63. *Caglioti, G.:* In: Thermal neutron diffraction, p. 14 (ed. *B. T. M. Willis*). London: Oxford University Press 1970.
64. *Nathans, R., Will, G., Cox, D. E.:* Proc. Intern. Conf. Magnetism, Nottingham 1964, 327.
65. *Collins, M. F., Tondon, V. K.:* Can. J. Phys. *50*, 2991 (1972).
66. *Anderson, P. W.:* Phys. Rev. *86*, 694 (1952).
67. *Davis, H. H.:* Phys. Rev. *120*, 789 (1960).
68. *Dzialoshinskii, I. E.:* Sov. Phys. JETP (Engl. Transl.) *6*, 1120 ,1259 (1957).
69. *Rietveld, H. M.:* J. Appl. Cryst. *2*, 65 (1969).
70. *Hewat, A. W.:* J. Phys. C: Solid State Phys. *6*, 2559 (1973).
71. *Greaves, C., Jacobson, A. J., Tofield, B. C., Fender, B. E. F.:* Acta Cryst. B, *31*, 641 (1975).
72. *Infante, C.:* personal communication 1974.
73. *Bonnet, M., Delapalme, A., Tchéou, F., Fuess, H.:* Polarized neutron determination of magnetic moments and magnetic form factors of $Fe^{3+}$ in yttrium iron garnet. Proc. Intern. Cong. Magnetism, Moscow 1973, IV, 251 (1974).
74. *Moon, R. M.:* Intern. J. Magnetism *1*, 219 (1971).
75. *Balcar, E., Lovesey, S. W., Wedgwood, F. A.:* J. Phys. C: Solid State Phys. *6*, 3746 (1973).
76. *Soules, T. F., Richardson, J. W.:* Phys. Rev. Letters *25*, 110 (1970).
77. *Freund, P.:* J. Phys. C: Solid State Phys. *7*, L33 (1974).
78. *Rimmer, D. E.:* Ref. 63, p. 211.
79. *Watson, R. E., Freeman, A. J.:* Acta Cryst. *14*, 27 (1961).
80. *Blume, M.:* Phys. Rev. *124*, 96 (1961).
81. *Freeman, A. J., Watson, R. E.:* Phys. Rev. *120*, 1125 (1960).
82. *Hutchings, M. T., Guggenheim, H. J.:* J. Phys. C: Solid State Phys. *3*, 1303 (1970).
83. *Shulman, R. G., Sugano, S.:* Phys. Rev. *130*, 506 (1963).

84. *Hall, T. P. P., Hayes, W., Stevenson, R. W. H., Wilkens, J.:* J. Chem. Phys. *38*, 1977 (1963); *39*, 35 (1963).
85. *Schoenberg, A., Suss, J. T., Szapiro, S., Luz, Z.:* Solid State Commun. *14*, 811 (1974).
86. *Rinneberg, H., Haas, H., Hartmann, H.:* J. Chem. Phys. *50*, 3064 (1969).
87. *Thornley, J. H. M., Windsor, C. G., Owen, J.:* Proc. Roy. Soc. (London) A *284*, 252 (1965).
88. *Thornley, J. H. M.:* D. Phil. Thesis, University of Oxford (1962).
89. *Windsor, C. G., Thornley, J. H. M., Griffiths, J. H. E., Owen, J.:* Proc. Phys. Soc. (London) *80*, 803 (1962).
90. *Sugano, S., Shulman, R. G.:* Phys. Rev. *130*, 517 (1963).
91. *Wachters, A. J. H., Nieuwpoort, W. C.:* Phys. Rev. B *5*, 4291 (1972).
92. *Ferguson, J.:* Progr. Inorg. Chem. *12*, 159 (1970).
93. *Brown, R. D., Burton, P. G.:* Theoret. Chim. Acta *18*, 309 (1970).
94. *Larsson, S., Connolly, J. W. D.:* J. Chem. Phys. *60*, 1514 (1974).
95. *Johnson, K. H.:* Intern. J. Quantum Chem. *1*, 361 (1967).
96. *Mott, N. F., Zinamon, Z.:* Rept. Progr. Phys. *33*, 881 (1970).
97. *Mattheiss, L. F.:* Phys. Rev. B *5*, 290, 306 (1972).
98. *Johnson, K. H., Messmer, R. P., Connolly, J. W. D.:* Solid State Commun. *12*, 313 (1973).
99. *Kuska, H. A., Rogers, M. T.:* J. Chem. Phys. *41*, 3802 (1964).
100. *Owen, J., Taylor, D. R.:* J. Appl. Phys. *39*, 791 (1968).
101. *Jacobson, A. J., McBride, L., Fender, B. E. F.:* J. Phys. C: Solid State Phys. *7*, 783 (1974).
102. *Freund, P., Owen, J., Hann, B. F.:* J. Phys. C: Solid State Phys. *6*, L 139 (1973).
103. *Davies, J. J., Smith, S. R. P., Owen, J., Hann, B. F.:* J. Phys. C: Solid State Phys, *5*, 245 (1972).
104. *Helmholz, L., Guzzo, A. V., Sanders, R. N.:* J. Chem. Phys. *35*, 1349 (1961).
105. *Shulman, R. G., Knox, K.:* Phys. Rev. Letters *4*, 603 (1960).
106. *Engelsman, F. M. R., Wiegers, G. A., Jellinek, F., Van Laar, B.:* J. Solid State Chem. *6*, 574 (1973).
107. *Van Laar, B., Ijdo, D. J. W.:* J. Solid State Chem. *3*, 590 (1971).
108. *Van Laar, B., Engelsman, F. M. R.:* J. Solid State Chem. *6*, 384 (1973).
109. *Ikeda, R., Nakamura, N., Kubo, M.:* J. Phys. Chem. *69*, 2101 (1965).
110. *Moss, J., Brown, P. J.:* J. Phys. F *2*, 358 (1972).
111. *Yamada, I., Kubo, H., Shimohigashi, K.:* J. Phys. Soc. Japan *30*, 896 (1971).
112. *Roth, W. L.:* J. Phys. Chem. Solids *25*, 1 (1964).
113. *Hastings, J. M., Elliott, N., Corliss, L. M.:* Phys. Rev. *115*, 13 (1959).
114. *Hazony, Y.:* Phys. Rev. B *3*, 711 (1971).
115. *Hewett, A. W.:* (1972) private communication, using the method of Ref. *(50)*.
116. *Freeman, A. J., Ellis, D. E.:* Phys. Rev. Letters *24*, 516 (1970).
117. *Lindgard, P. A., Marshall, W.:* J. Phys. C: Solid State Phys. *2*, 276 (1969).
118. *McLachlan, A. D.:* Mol. Phys. *3*, 233 (1960).
119. *Nathans, R., Pickart, S. J., Alperin, H. A.:* Bull. Am. Phys. Soc. *5*, 455 (1960).
120. *Shull, C. G., Strausar, W. A., Wollan, E. O.:* Phys. Rev. *81*, 333 (1951).
121. *Roth, W. L.:* Phys. Rev. *111*, 772 (1958).
122. *Van Laar, B.,* Phys. Rev. *138*, A 584 (1965), *Van Laar, B., Schweizer, J., Lemaire, R.:* Phys. Rev. *141*, 538 (1966).
123. *Khan, D. C., Erickson, R. A.:* Phys. Rev. B *1*, 2243 (1970).
124. *Mahendra, A., Khan, D. C.:* Phys. Rev. B *4*, 3901 (1971).
125. *Kanamori, J.:* Progr. Theoret. Phys. (Kyoto) *17*, 177 (1957).
126. *Lovesey, S. W., Rimmer, D. E.:* Rept. Progr. Phys. *32*, 333 (1969).
127. *Fender, B. E. F., Coffin, P. S.:* 1971 (unpublished).
128. *Walker, M. B., Stevenson, R. W. H.:* Proc. Phys. Soc. (London) *87*, 35 (1966).
129. *Helmholz, L., Guzzo, A. V.:* J. Chem. Phys. *32*, 302 (1960).
130. *Jacobson, A. J., Fender, B. E. F.:* J. Phys. C: Solid State Phys., *8*, 844 (1975).
131. *Hannay, N. B., Smyth, C. P.:* J. Am. Chem. Soc. *68*, 171 (1946).
132. *Matumura, O.:* J. Phys. Soc. Japan *14*, 108 (1959).
133. *Watson, R. E., Freeman, A. J.:* Phys. Rev. *123*, 2027 (1961).

134. *Geschwind, G.:* In: Hyperfine interactions p. 225 (ed. *A. J. Freeman* and *R. B. Frankel*). New York: Academic Press 1967.
135. *Rinneberg, H.H., Shirley, D.A.:* Phys. Rev. Letters *30*, 1147 (1973).
136. *Huang, N.L., Orbach, R., Simánek, E., Owen, J., Taylor, D. R.:* Phys. Rev. *156*, 383 (1967).
137. *Evans, B. J., Swartzendruber, L. J.:* Phys. Rev. B *6*, 223 (1972).
138. *Lau, B. F., Newman, D. J.:* J. Phys. C: Solid State Phys. *6*, 3245 (1973).
139. *Shannon, R. D., Prewitt, C. T.:* Acta Cryst. B *25*, 925 (1969).
140. *Wollan, E. O., Koehler, W. C.:* Phys. Rev. *100*, 545 (1955).
141. *Goodenough, J. B., Wold, A., Arnott, R. J., Menyuk, N.:* Phys. Rev. *124*, 373 (1961).
142. *Wollan, E. O., Child, H. R., Koehler, W. C., Wilkinson, M. K.:* Phys. Rev. *112*, 1132(1958).
143. *Tofield, B. C.:* unpublished work.
144. *Wilkinson, M. K., Wollan, E. O., Child, H. R., Cable, J. W.:* Phys. Rev. *121*, 74 (1961).
145. *Baker, J. M., Blake, W. B. J.:* Phys. Letters A *31*, 61 (1970).
146. *Newman, D. J.:* Advan. Phys. *20*, 197 (1971).
147. *Brun, T. O., Lander, G. H.:* Phys. Rev. B *9*, 3003 (1974).
148. *Freeman, A. J., Watson, R. E.:* Phys. Rev. *127*, 2058 (1962).
149. *Blume, M., Freeman, A. J., Watson, R. E.:* J. Chem. Phys. *37*, 1245 (1962); *41*, 1878 (1964).
150. *Freeman, A. J., Desclaux, J. P.:* Intern. J. Magnetism *3*, 311 (1972).
151. *Lander, G. H., Brun, T. O., Desclaux, J. P., Freeman, A. J.:* Phys. Rev. B *8*, 3237 (1973).
152. *Moon, R. M.:* (1974), personal communication.
153. *Wagner, F. E., Thoma, K., Campbell, L. E., Kalvius, G. M.:* The observation of a magnetic transition in the perovskite $BaPrO_3$ by means of the Mössbauer effect. Proc. Intern. Conf. Magnetism, Moscow (1973).
154. *Baker, J. M., Chadwick, J. R., Garton, G., Hurrell, J. P.:* Proc. Roy. Soc. (London) A *286*, 352 (1965).
155. *Watson, R. E., Freeman, A. J.:* Phys. Rev. *156*, 251 (1967).
156. *Freeman, A. J.:* (1974), personal communication.

Received July 31, 1974

# Superheavy Elements

## A Prediction of Their Chemical and Physical Properties

**Burkhard Fricke**

Gesamthochschule Kassel, D-3500 Kassel, Heinrich-Plett-Str. 40, Germany and
Gesellschaft für Schwerionenforschung, D-6100 Darmstadt, Postfach 541, Germany

## Table of Contents

# I. Introduction

Very recently *Oganesian, Flerov* and coworkers (*1*) in Dubna announced the discovery of element 106. Although they observed less than 100 fission tracks of the decaying nuclei of this element, formed after the heavy-ion bombardment of Cr ions on Pb, they were able to measure a half-life of about 20 msec for one isotope and 7 msec for another. At about the same time *Ghiorso* and coworkers (*2*) in Berkeley found a new alpha activity for which they established the genetic link with the previously identified daughter and grand-daughter nuclides

$$^{263}106 \xrightarrow[0.9\,\text{sec}]{\alpha} \ ^{259}104 \xrightarrow[3\,\text{sec}]{\alpha} \ ^{255}102 \xrightarrow[3\,\text{min}]{\alpha} \ .$$

This evidence indicates that a new element has been added to the periodic table, thus presenting a new challenge to scientists.

Until 1940 the heaviest known element was uranium with the atomic number 92, and at that time (about 1944) the actinide concept of *Seaborg* (*3*) was just a hypothesis. It is thus apparent that great progress has been made since then. Fourteen new elements have been added to the periodic system and much chemical and physical information has been gathered concerning this region of elements. Hence we can expect that element 106 is probably not the last element but only a step toward an even longer periodic table. The approach used in the experiments up to now to produce even heavier transuranium elements has been to proceed element by element into the region of atomic numbers just beyond the heaviest known by bombarding high-$Z$ atoms with small-$Z$ atoms. There have been very difficult and laborious attempts to proceed even further (*4, 5*). The upper limit of this method is determined by experimental feasibility; it cannot now be predicted with certainty but will be about element 108 or 109. The other way to proceed is to bombard two very heavy elements with each other, thus producing superheavy elements directly. This method will probably overlap with the first method at its lower end.

This second method, which must be the result of bombardments with relatively high $Z$ heavy ions, is still in preparation at several places in the world, *i. e.* Dubna in the USSR, Berkeley in the USA, Orsay in France, and Darmstadt in Germany. If this method is successful, it should lead to the nearly simultaneous discovery of a number of new elements.

There is general agreement that theoretical predictions of nuclear stability, which we discuss briefly in the next paragraph, define a range of superheavy elements with sufficiently long half-lives to allow their study, provided they can be synthesized. What cannot be predicted is whether there exist nuclear reactions for such synthesis in detectable amounts on earth.

The known elements heavier than uranium are usually called by the very unspecific name of transuranium elements. In the upper range this term is expected to overlap with the equally ill-defined expression superheavy elements. To clarify the situation from a nuclear physics point of view, one may define the end of the transuranium elements and the beginning of the superheavy elements as the element where the nuclear stability of the longest-lived isotope increases again with increasing $Z$. The observed strong decrease of the half-lives of the transuranium elements known up to now can be seen in Fig. 1. The question is where and if this trend to even smaller half-lives is likely to end.

From a chemical point of view the elements, including the unknown superheavy elements, are well defined by their location in the periodic table. The elements up to 103 are the actinides or the $5f$ transition elements. Chemical reviews of these are given by *Seaborg (6, 7)*, *Cunningham (8)*, *Asprey* and *Pennemann (9)*, and *Keller (10)*. The $6d$ transition series starts with element 104. Of course, the first chemical question to be answered is whether this simple series concept of the periodic table still holds for the superheavy elements. A very comprehensive review of elements 101 to 105, discussing the nuclear stability and chemical behavior of the predicted elements, was given by Seaborg (5) in 1968. Several other articles dealing mainly with the chemical behavior of superheavy elements, the search for superheavy elements in nature, and the electronic structure of these elements have since been published. The references are given in the discussion below.

In this summary of the very quickly developing field of the superheavy elements, the main emphasis lies on the prediction of their chemical properties. Apart from the general interest of the question, this knowledge is expected to be very important because chemical separation will be one of the methods used to detect superheavy elements.

## II. Predictions of Nuclear Stability

Like the well-known effect of the closed shells in the atomic electron cloud at $Z = 2, 10, 18, 36, 54, 86$, which is the physical basis for the structure of the periodic table, the effect of closed nucleon shells together with a large separation to the next unoccupied shell also makes for considerable nuclear stability. The nucleus consists of two kinds of particles, protons and neutrons, so that we have two series of so-called magic numbers. These are for protons 2, 8, 20, 28, 50, 82, and for neutrons 2, 8, 20, 28, 50, 82, and 126. Nuclei where both protons and neutrons are magic, ($^{16}O$, $^{40}Ca$ or $^{208}Pb$, for example) are called double-magic nuclei and are particularly stable. As we go to even heavier nuclei, the effect which most heavily influences stability to $\alpha$-decay or fission (the most important decay modes) is the increasingly large repulsion of the nucleonic charges against the attractive nuclear forces, which severely shortens the half-lives of the nuclei (*11*), as can be seen from Fig. 1. This suggests the question: Is the stabilizing effect of the next

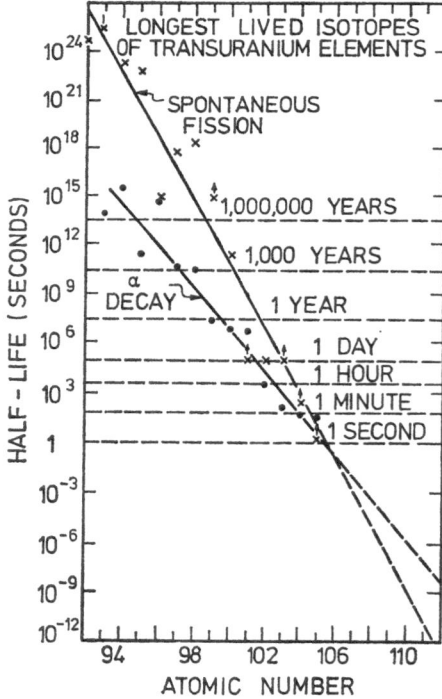

Fig. 1. The longest-lived isotopes of transuranium elements as a function of $Z$ for spontaneous fission and $\alpha$ decay (*11*)

double-magic configuration large enough to counteract this repulsion and to lengthen the half-lives yet again?

Because it was assumed that the next protonic magic number was 126 (by analogy with the neutrons), early studies of possible superheavy elements did not receive much attention (12—15), since the predicted region was too far away to be reached with the nuclear reactions available at that time. Moreover, the existence of such nuclei in nature was not then considered possible. The situation changed in 1966 when *Meldner* and *Röper* (16, 17) predicted that the next proton shell closure would occur at atomic number 114, and when *Myers* and *Swiatecki* (18) estimated that the stability fission of a superheavy nucleus with closed proton and neutron shells might be comparable to or even higher than that of many heavy nuclei.

These results stimulated extensive theoretical studies on the nuclear properties of superheavy elements (19). The calculations published so far have been based on a variety of approaches. Most calculations were performed by using a phenomenological description within the deformed shell model (20—23). In this model the nucleons are considered to move in an average potential and the shape of the potential and other parameters are chosen by fitting single-particle levels in well-investigated spherical or deformed nuclei. Regardless of the approach followed, the authors agree in predicting a double-magic nucleus $^{298}114$, although several other magic proton and neutron numbers near these values have been discussed.

There are also several self-consistent calculations (17, 24—27) but suitable parameters have to be used, because the nucleon-nucleon force is not known from general considerations. Most authors also accept the magic numbers $Z = 114$ and $N = 184$.

In addition to the proton magic number 114, a second superheavy magic proton number was investigated at $Z = 164$ (23, 28). Although the realization of such a nucleus seems to be far from any practical possibility at the moment, one should bear this region in mind because many most interesting questions could be answered if it were possible to produce these elements. One way to actually proceed

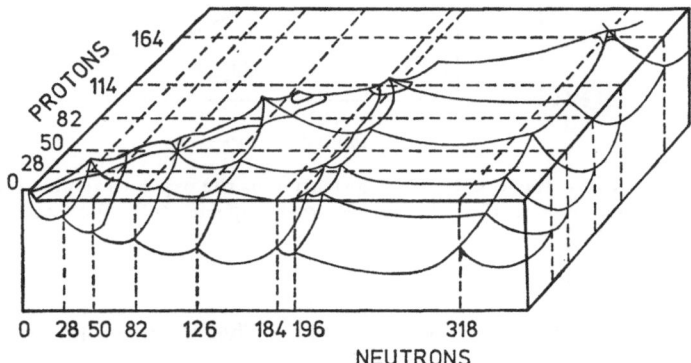

Fig. 2. Schematic drawing of the stability of the nuclei as a function of the number of protons and neutrons. The expected islands of stability can be seen near $Z = 114$ and $Z = 164$ (29)

into this region is the observation of the X-rays from the quasi-molecular systems which are transiently formed during heavy ion collission (109).

These predictions are depicted very schematically in Fig. 2 in an allegorical fashion (29). The long peninsula corresponds to the region of known nuclei. The grid lines represent the magic numbers of protons (Z) and neutrons (N). The third dimension represents the stability. The magic numbers are shown as ridges and the double-magic nuclei, like $^{208}$Pb, are represented as mountains. The two regions near $Z = 114$ and $Z = 164$ show up in Fig. 2 as "islands of stability" within the large "sea of instability".

The detailed calculations quoted above predict potential barriers against fission, i.e. the total energy of the nucleus is calculated as a function of the deformation, because a deformation parameter describes at the one extreme the spherical nucleus and at the other the two separated nuclei after fission. All of these calculations indicate a maximum (or two) at a small deformation, whereas we get a dip of a few MeV at zero deformation and a trough of a few hundred MeV for very large deformations. The result of such a calculation is shown in Fig. 3

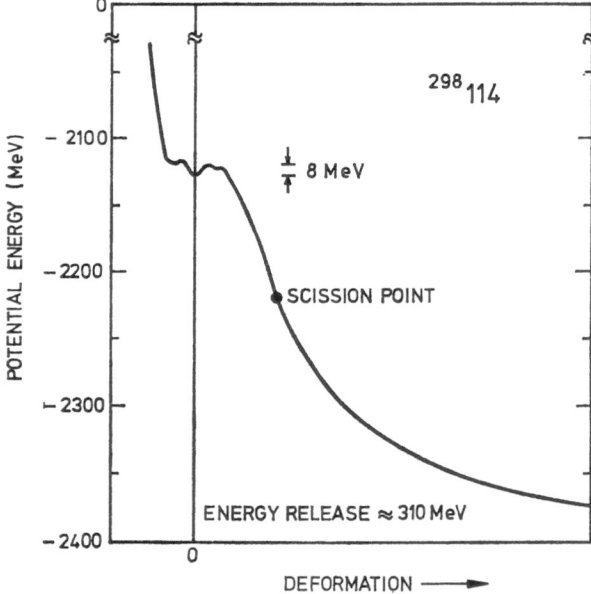

Fig. 3. Total energy as a function of the deformation of the expected double-magic nucleus $^{298}$114. The small minimum at the deformation zero is expected to be the reason for the very long lifetime of this nucleus (32)

for the expected double-magic nucleus $^{298}$114. This small minimum at zero deformation plays an important role; it keeps the nucleus in spherical shape and prevents rapid decay in the fission path. Spontaneous fission can occur only by the extremely slow process of tunneling through the several MeV high barrier.

Thus, the height and width of this barrier play a most important role in the prediction of the half-lives against fission (*31*). For the double-magic nucleus $^{298}114$ a height of between 9 and 14 MeV is predicted, depending on the method used. This yields spontaneous fission half-lives of between $10^7$ and $10^{15}$ years.

These first results were very promising and stimulated a very extensive but up to now unsuccessful search for superheavy elements in nature. A most comprehensive review of this subject was given by *G. Herrmann* (*32*). But, besides spontaneous fission, a nucleus can decay by other decay modes like $\alpha$ decay, $\beta$ decay, or electron capture. The most comprehensive study of half-lives in the first superheavy island was performed by *Fiset* and *Nix* (*33*). Figure 4 is taken from their work.

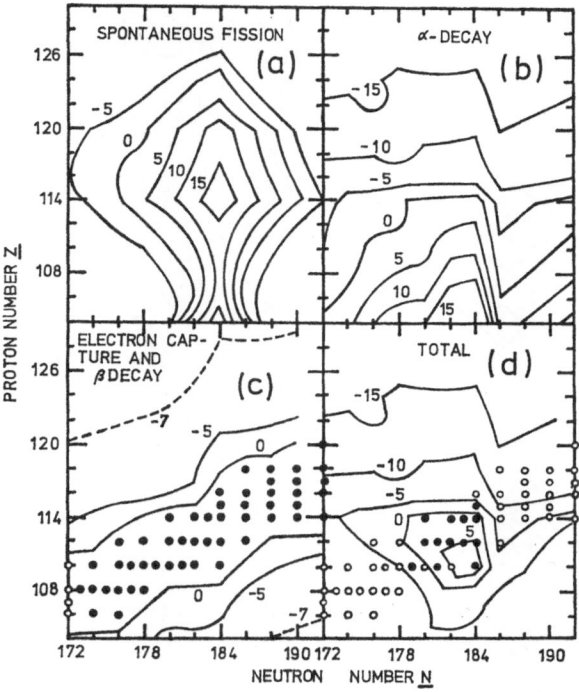

Fig. 4. Summary of predictions of the half-lives of the nuclei at the first island of stability. (a) spontaneous-fission half-lives, (b) $\alpha$-decay half-lives, (c) electron-capture and $\beta$-decay half-lives, and (d) total half-lives. The numbers give the exponent of 10 of the half-lives in years (*33*)

The results show that, as one moves away from the double closed-shell nucleus $^{298}114$, the calculated spontaneous fission half-lives in Fig. 4a decrease from $10^{15}$ y for nuclei on the inner contour to $10^{-5}$ y (about 5 min) for nuclei on the outer contour. With respect to spontaneous fission, the island of superheavy nuclei is a mountain ridge running north and south, with the descent being most gentle in the northwest direction. The calculated $\alpha$-decay half-lives in Fig. 4b, however, decrease, rather smoothly with increasing proton number from $10^5$ y for nuclei along the bottom contour to $10^{-15}$ y (about 30 nsec) for nuclei along the top contour. The discontinuities arise from shell effects. The $\beta$-stability valley crosses the island from the southwest to the northeast direction.

The calculated $\beta$-decay and electron-capture half-lives in Fig. 4c decrease from 1 y for nuclei along the inner contour to $10^{-7}$ y (about 3 sec) for nuclei at the outer contour. The total half-lives in Fig. 4d are obtained by taking into account all three decay modes. The longest total half-life of $10^9$ years is found for the nucleus $^{294}110$. A three-dimensional plot of these results (35) is given in Fig. 5, where the island character of this region of relative stability is beautifully demonstrated.

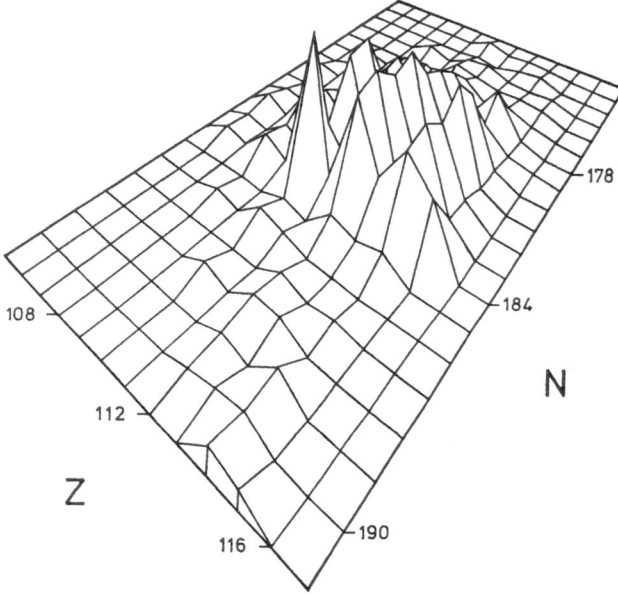

Fig. 5. Same as Fig. 4d in a three-dimensional plot (35)

In considering such results (34), one should be aware of the great uncertainties associated with the extrapolation of nuclear properties into the unknown region. The calculations are associated with large errors. The total uncertainty for all three decay modes discussed here is as large as $10^{10}$ for the half-lives.

The half-lives for the second island of stability are even smaller. In the most optimistic estimates, they are not more than a few hours, and the uncertainty of $10^{10}$ brings them down into the region of nsec. The second assumption of course, which has still to be proved, is that these nuclei can in fact be produced.

In conclusion, one may say that there is general agreement that theoretical predictions of nuclear stability define a range of superheavy elements in the vicinity of element 114 with sufficiently long half-lives to allow their study, provided they can be synthesized.

# III. Basis for the Predictions of Chemical and Physical Properties

## 1. The Electronic Structure

When *Mendeleev* constructed his periodic system in 1869 he had actually found the most general and overall systematics known in science. He developed this table from his comparison of the chemical and physical properties of the elements, without knowing the underlying reason for it. Since the early stages of quantum mechanics in the 1920's, it has become clear that the similarity of the properties of the elements depends strongly on the outer electronic structure. The filled-shell concept is in accord with the periodicity of the chemical properties that formed the basis for the concept of the periodic table.

Thus it is obvious that the first step toward predicting chemical and physical properties is to predict the electronic structure of the superheavy elements. An excellent review article on this subject will be published by *J. B. Mann* (35) in the near future.

**a) Continuation of the Periodic Table.** As early as 1926 *Madelung* (36) found the empirical rules for the electron-shell filling of the ground-state configurations of the neutral atoms. His rules are simple:

1. electron shells fill in order of increasing value of the quantum number sum $(n + l)$, where $n$ is the principal quantum number and $l$ the orbital quantum number;

2. for fixed $(n + l)$, shells fill in order of increasing $n$.

In Fig. 6 we show one of the many published schemes based on these rules, which demonstrates the filling of the electrons. This systematics provides an almost correct explanation of all known neutral atomic configurations in the known region of elements. This simple law was therefore used by *Gol'danskii* (37) and by *Seaborg* (5) to predict the electron structure of the superheavy elements. *Seaborg* designated the 32 elements of the 5g and 6f shells as the "superactinide series" and placed them as elements 122 to 153 by analogy with the actinide series

|  |  |  |  |  |  | $(n+l)$ |
|---|---|---|---|---|---|---|
|  |  |  |  | 1 1s 2 | | 1 |
|  |  |  |  | 3 2s 4 | | 2 |
|  |  |  | 5 2p 10 | 11 3s 12 | | 3 |
|  |  |  | 13 3p 18 | 19 4s 20 | | 4 |
|  |  | 21 3d 30 | 31 4p 36 | 37 5s 38 | | 5 |
|  |  | 39 4d 48 | 49 5p 54 | 55 6s 56 | | 6 |
|  | 57 4f 70 | 71 5d 80 | 81 6p 86 | 87 7s 88 | | 7 |
|  | 89 5f 102 | 103 6d 112 | 113 7p 118 | 119 8s 120 | | 8 |
| 121 5g | 6f | 7d | 8p 168 | 169 9s 170 | | 9 |

Fig. 6. The filling of the electron shells according to the simple rule of Madelung (*36*)

90 to 103 following actinium. Since there were already in the known region of elements a few deviations from *Madelung's* simple rules, especially in the lanthanides and actinides, *Chaikkorskii* (38) and later *Taube* (39) tried to predict these anticipated deviations. In Table 1 we show the predictions of *Gol'danskii, Seaborg,*

Table 1. Predictions of the ground-state configurations of *Gol'danskii* (37), *Chaikkorskii* (38), *Taube* (39) and *Seaborg* (5) for elements 121 to 127 and 159 to 168, using the principle of the extrapolation within the periodic table. The main quantum numbers (5g, 6f, 7d, 8s) are not shown. This table is taken from *Mann* (35)

| Element | Taube (39) | Gol'danskii (37) | Seaborg (57) | Chaikkorskii (38) |
|---------|------------|------------------|--------------|-------------------|
| 121 | $ds^2$ | $gs^2$ | $ds^2$ | $ds^2$ |
| 122 | $fds^2$ | $g^2 s^2$ | $g^2 s^2$ | $d^2 s^2$ |
| 123 | $gfds^2$ | $g^3 s^2$ | $g^3 s^2$ | $d^3 s^2$ |
| 124 | $g^2 fds^2$ | $g^4 s^2$ | $g^4 s^2$ | $f^2 d^2 s^2$ |
| 125 | $g^3 fds^2$ | $g^5 s^2$ | $g^5 s^2$ | $g^2 f^2 ds^2$ |
| 126 | $g^4 fds^2$ | $g^6 s^2$ | $g^6 s^2$ | $g^3 f^2 ds^2$ |
| 127 | $g^5 fds^2$ | $g^7 s^2$ | $g^7 s^2$ | $g^5 d^2 s^2$ |
| 159 | $d^7 s^2$ | $d^7 s^2$ | | $d^7 s^2$ |
| 160 | $d^8 s^2$ | $d^8 s^2$ | | $d^9 s^1$ |
| 161 | $d^9 s^2$ | $d^9 s^2$ | | $d^{10} s^1$ |
| 162 | $d^{10} s^2$ | $d^{10} s^2$ | | $d^{10} s^2$ |
| 163—168 | $8s^2 8p^n$ ($n = 1 - 6$) for all columns | | | |

*Chaikkorskii* and *Taube* for elements 121 to 127 and 159 to 168. Apart from small discrepancies in these somewhat uncertain regions, there was general agreement that the unfinished 8th row of the periodic table would be finished by the 6d elements ending at element 112 and the 7p elements at 118. From a conservative point of view, every extrapolation into the region starting with element 121 is expected to be very speculative. Nevertheless, the reliability of the location of the elements in the periodic table seems to be relatively unambiguous.

**b) Ab-initio Atomic Calculations.** The prediction of the electronic configurations of the superheavy elements became much more reliable when ab-initio atomic calculations became available and accurate enough to be used in the field of the superheavy elements.

In the following paragraph we give a very brief description of the principles used in the calculations. For the details, especially the exact formulas used, we refer to the literature. All the calculations that are useful in this connection are based on the calculation of the total energy $E_T$ of the electronic system, given by the expression

$$E_T = \frac{\langle \psi \,|\, H \,|\, \psi \rangle}{\langle \psi \,|\, \psi \rangle}$$

where $\psi$ is the total wave function and $\boldsymbol{H}$ the Hamiltonian of the system. The physical solution is found when $E_T$ is at the total minimum after the variation of $\psi$.

Depending on the ansatz for the total wave functions $\psi$ and the Hamiltonian $\boldsymbol{H}$ of the system, this minimalization of the total energy leads to a set of different, usually coupled differential equations. The solution of these equations gives the total wave function and hence the total energy. These methods, usually called Hartree and Hartree-Fock methods, are described in detail in various texts and papers (40). For those planning to do such calculations, R. D. Hartree's "Calculation of Atomic Structure" (41), and "Atomic Structure Calculations" (42) by F. Herrmann and S. Skillman are recommended. A review article by I. P. Grant (43) gives an excellent description of relativistic methods. A good summary is also given by J. B. Mann (35).

Let us discuss very briefly the various methods that have been used. The first group of calculations is done by using the non-relativistic Hamiltonian (ignoring spin-orbit interaction)

$$\boldsymbol{H} = -\sum_i \frac{1}{2} \nabla^2 - \sum_i \frac{Z}{r_i} + \sum_{i<j} \frac{1}{r_{ij}}.$$

Here the first term with $\nabla$ the Nabla operator is the kinetic energy, the second is the potential energy due to the nuclear charge, and the last term is the total electrostatic interaction energy over all pairs of electrons.

*Hartree's method* (H) considers the total wave function to be a product of one-electron wave functions $\psi = \prod_i \varphi_i$; this leads, after the variation of the total energy, to a set of second-order homogeneous differential equations that have to be solved for the radial wave functions of the electrons of each shell. The last term due to the interaction of the electrons is given by the potential generated by all the other electrons. In this respect the set of the differential equations is, of course, already coupled. This was the basic method used by Larson et al. (44) for the first atomic claculations in the region of superheavy elements $Z = 122$ to 127.

*Hartree-Fock method* (HF). Here the total wave function is assumed to be an antisymmetric sum of Hartree functions and can be represented by a Slater determinant

$$\psi = (N!)^{-\frac{1}{2}} |\varphi_1(1)\, \varphi_1(2) \,.....\, \varphi_N(N)|$$

which automatically obeys the Pauli principle.

The effect of the determinantal wave function is to greatly complicate the resulting differential equations by adding exchange potential terms, giving rise to an inhomogeneous equation, for which the correct solutions becomes much more difficult and time-consuming. For the exact equations, see for example J. B. Mann (35). A program using this method was developed by C. Froese-Fischer (45).

*Hartree-Fock-Slater method* (HFS). In this method the inhomogeneous parts of the equations used in the Hartree-Fock method are approximated by a local potential, as proposed in 1951 by *Slater* (46). This approximation yields much simpler homogeneous differential equations in which the potential terms are identical for every orbital of the atom, which makes the actual computation less time-consuming by a factor of about 5 to 10, although the results are nearly as good as with the Hartree-Fock method.

For heavy elements, all of the above non-relativistic methods become increasingly in error with increasing nuclear charge. *Dirac* (47) developed a relativistic Hamiltonian that is exact for a one-electron atom. It includes relativistic mass-velocity effects, an effect named after Darwin, and the very important interaction that arises between the magnetic moments of spin and orbital motion of the electron (called spin-orbit interaction). A completely correct form of the relativistic Hamiltonian for a many-electron atom has not yet been found. However, excellent results can be obtained by simply adding an electrostatic interaction potential of the form used in the non-relativistic method. This relativistic Hamiltonian has the form

$$ \boldsymbol{H} = \sum_k \left( i\, c\, \alpha(k) \cdot \nabla(k) - \beta(k)\, c^2 - \frac{Z}{r_k} \right) + \sum_{k<j} \frac{1}{r_{ij}} , $$

where $\alpha$ and $\beta$ are $4 \times 4$ matrices and $\nabla$ is the Nabla operator. Using the variational method in the same manner as before and taking a Slater determinant as the wave function, one obtains two sets of first-order inhomogeneous differential equations to be solved for all electrons of the atom. This most complicated version of atomic calculations is called the *relativistic Hartree-Fock method* or *Dirac-Fock method* (DF) (48). Various papers calculating the ground-state configurations of superheavy atoms by this method have been published since 1969 (49—51). A complete discussion is given by *J. B. Mann* (35).

These very complicated inhomogeneous coupled differential equations can again be simplified by using Slater's approximation. This method is therefore called the *relativistic Hartree-Fock-Slater* or *Dirac-Fock-Slater* (DFS) (52—53) calculations, and they have also been done by several authors for the superheavy elements (54—56).

The results for the ground-state configurations of all superheavy elements up to 172 and for element 184 are given in Table 2 (35, 50, 56—60). In only very few cases are the results different for the two best methods, DF and DFS, but the differences are so small that no final decision can be made.

The first difference that becomes obvious in comparison to the empirical continuation of the electron filling discussed above (29, 37—39) occurs at elements 110 and 111. The calculated ground-state is $s^2 d^8$ and $s^2 d^9$, respectively, which is not at all common in the homologs of the two elements.

Also, beginning with element 121, every element has a different ground-state configuration than that predicted by simple extrapolations. The main reason for this behavior is that, unexpectedly, an 8p electron state becomes occupied at element 121, and at least one of these electrons remains bound through all the following elements. In the 160 region the difference between the simple predictions and the results of the calculations is already so large that the position of the elements in the periodic table is changed drastically. (For an overview and com-

Table 2. Atomic ground-state configurations for the neutral elements 103 to 172 and 184 according to *Mann* (35) and *Fricke* and *Waber* (85, 60), using self-consistent Dirac–Fock calculations

| Element | Rn core + $5f^{14}$ + | Element | $Z = 120$ core + | Element | $Z = 120$ core + $8p^2_{1/2}$ + | Element | $Z = 120$ core + $8p^2_{1/2}\ 5g^{18}\ 6f^{14}$ + |
|---|---|---|---|---|---|---|---|
| 103 | $7s^2\ 7p$ | 121 | $8p$ | 139 | $5g^{13}\ 6f^2\ 7d^2$ | 157 | $7d^3$ |
| 104 | $7s^2\ 6d^2$ | 122 | $8p\ 7d$ | 140 | $5g^{14}\ 6f^3\ 7d^1$ | 158 | $7d^4$ |
| 105 | $7s^2\ 6d^3$ | 123 | $8p\ 7d\ 6f$ | 141 | $5g^{15}\ 6f^2\ 7d^2$ | 159 | $7d^4\ 9s^1$ |
| 106 | $7s^2\ 6d^4$ | 124 | $8p\ 6f^3$ | 142 | $5g^{16}\ 6f^2\ 7d^2$ | 160 | $7d^5\ 9s^1$ |
| 107 | $7s^2\ 6d^5$ | 125 | $8p\ 6f^3\ 5g$ | 143 | $5g^{17}\ 6f^2\ 7d^2$ | 161 | $7d^6\ 9s^1$ |
| 108 | $7s^2\ 6d^6$ | 126 | $8p\ 7d\ 6f^2\ 5g^2$ | 144 | $5g^{18}\ 6f^1\ 7d^3$ | 162 | $7d^8$ |
| 109 | $7s^2\ 6d^7$ | 127 | $8p^2\ 6f^2\ 5g^3$ | 145 | $5g^{18}\ 6f^3\ 7d^2$ | 163 | $7d^9$ |
| 110 | $7s^2\ 6d^8$ | 128 | $8p^2\ 6f^2\ 5g^4$ | 146 | $6f^4\ 7d^2$ | 164 | $7d^{10}$ |
| 111 | $7s^2\ 6d^9$ | 129 | $8p^2\ 6f^2\ 5g^5$ | 147 | $6f^5\ 7d^2$ | 165 | $7d^{10}\ 9s^1$ |
| 112 | $7s^2\ 6d^{10}$ | 130 | $8p^2\ 6f^2\ 5g^6$ | 148 | $6f^6\ 7d^2$ | 166 | $7d^{10}\ 9s^2$ |
| 113 | $7s^2\ 6d^{10}\ 7p$ | 131 | $8p^2\ 6f^2\ 5g^7$ | 149 | $6f^6\ 7d^3$ | 167 | $7d^{10}\ 9s^2\ 9p^1_{1/2}$ |
| 114 | $7s^2\ 6d^{10}\ 7p^2$ | 132 | $8p^2\ 6f^2\ 5g^8$ | 150 | $6f^6\ 7d^4$ | 168 | $7d^{10}\ 9s^2\ 9p^2_{1/2}$ |
| 115 | $7s^2\ 6d^{10}\ 7p^3$ | 133 | $8p^2\ 6f^3\ 5g^8$ | 151 | $6f^8\ 7d^3$ | 169 | $7d^{10}\ 9s^2\ 9p^2_{1/2}\ 8p^1_{3/2}$ |
| 116 | $7s^2\ 6d^{10}\ 7p^4$ | 134 | $8p^2\ 6f^4\ 5g^8$ | 152 | $6f^9\ 7d^3$ | 170 | $7d^{10}\ 9s^2\ 9p^2_{1/2}\ 8p^2_{3/2}$ |
| 117 | $7s^2\ 6d^{10}\ 7p^5$ | 135 | $8p^2\ 6f^4\ 5g^9$ | 153 | $6f^{11}\ 7d^2$ | 171 | $7d^{10}\ 9s^2\ 9p^2_{1/2}\ 8p^3_{3/2}$ |
| 118 | $7s^2\ 6d^{10}\ 7p^6$ | 136 | $8p^2\ 6f^4\ 5g^{10}$ | 154 | $6f^{12}\ 7d^2$ | 172 | $7d^{10}\ 9s^2\ 9p^2_{1/2}\ 8p^4_{3/2}$ |
| 119 | $7s^2\ 6d^{10}\ 7p^6\ 8s$ | 137 | $8p^2\ 6f^3\ 7d\ 5g^{11}$ | 155 | $6f^{13}\ 7d_5$ | | $Z = 172$ core + |
| 120 | $7s^2\ 6d^{10}\ 7p^6\ 8s^2$ | 138 | $8p^2\ 6f^3\ 7d\ 5g^{12}$ | 156 | $6f^{14}\ 7d^2$ | 184 | $6g^5\ 7f^4\ 8d^3$ |

parison, see the periodic table in Fig. 21, incorporating the results of the prediction of the elements up to 172 taken from *Fricke et al. (56)*). This disagreement with the results expected from a simple continuation of the periodic table is of course, a result of the interpretation of the periodic system in terms of chemical behavior, but the primary reason is the surprising order of filling of the outer electron shells in this region.

If we try to proceed to even heavier elements, the calculations come to a halt at $Z = 174$ because at this element the 1s level reaches the negative continuum of the electrons at a binding energy of $2m_e c^2 = 1$ MeV and the calculation breaks down. To proceed further, *Fricke (61)* introduced a phenomenological description of the quantum-electrodynamical effects into the SCF calculations *(60)*, which shifts the binding energies of the inner electrons back to lower values. Using this method, he was then able to study the electron configurations of the elements beyond $Z = 174$.

## 2. Trends of the Chemical and Physical Properties

The detailed and sophisticated calculations of the electronic-ground states of the atoms are very worthwhile as an important, though only the first step toward predicting the chemical and physical properties of superheavy elements, because chemistry consists not only of the properties of the atoms but also of the molecules and their behavior. Ab-initio calculations of molecules were introduced for small molecules and small $Z$, and the state of the art is still far away from the point that allows actual calculations of the chemical properties of superheavy molecules. A first step in this direction has been taken by *Averill et al. (62)*, who calculated the wave function of $(110)F_6$ using a muffin-tin method.

**a) Trends Emerging from the Calculations.** Although we are not able to calculate the properties of superheavy molecules at the present time, the atomic calculations give us more than just the electronic structure of the neutral elements.

One has to bear in mind that two elements from the same chemical group, which often have the same outer electronic structure will be chemically and physically slightly different. This can be to some extent explained as the effect of their somewhat different sizes, changed ionization potential, and the different energies and radial distributions of the wave functions between analogous shells. These quantities are also determined directly by the atomic calculations. The size of the atom or ion correlates strongly with the principal maximum of the outermost electronic shell, as found by *Slater (63)*, thus giving a first estimate of this important magnitude. Sometimes the expectation value of $<r>$ of the outermost shell is used as the radius, but the agreement with experiment is not so good.

There is considerable agreement that the ionization potentials have to be calculated in the adiabatic approximation, in which it is assumed that during the removal of an electron sufficient time elapses for the other electrons to rearrange themselves, so that the ionization potential is given by the difference in total energy of the two calculations with $m$ and $m-1$ electrons. The other method, taking the calculated energy eigenvalues *(64)*, can only be used as an approximation to this physical quantity.

In the first part of the periodic table it is relatively easy to make the connection between these quantities and a chemical interpretation because of the few shells involved and their large separation. Moreover, the influence of the inner electron shells is rather small so that the outer electron configurations are very similar in the same chemical group at different periods. When we proceed to higher elements at the end of the periodic table, the number of shells increases, the binding energy of the last electrons decreases, and there is competition between shells; hence the influence of the inner electrons becomes more significant. This rather complex behavior is further complicated by the fact that relativistic effects now begin to be important and the coupling between the angular momenta of the electrons changes from LS to intermediate or $j$–$j$ coupling. All these effects and their relative influences are taken into account in the ab-initio calculations. Of course, to prove their reliability in the superheavy region of elements, they have to reproduce the complex structure and its relationship to chemical behavior in the known part of the periodic table, which in all cases is done for example for the groundstate configurations of the atoms. The main change due to relativistic effects is the splitting of all shells with $l \neq 0$ into two subshells with $j = l + 1/2$ and $j = l - 1/2$. This means that, for example the $p$ state splits into the $p_{1/2}$ subshell

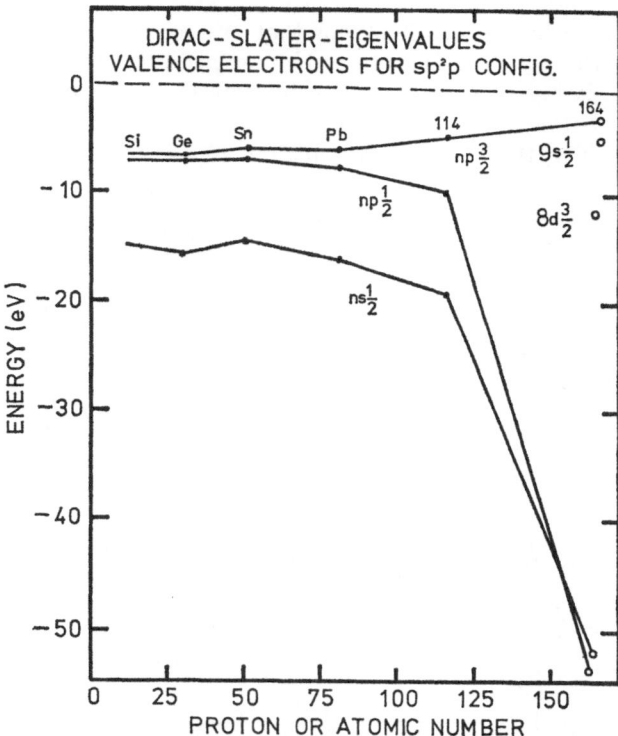

Fig. 7. Comparison of the eigenvalues of the $ns$, $np_{1/2}$ and $np_{3/2}$ electrons in the group-IVA elements using DFS calculations. This figure illustrates the very strong dependence of the spin-orbit splitting between the two $p$ states as a function of the atomic number. For element 164, the $9s$ and $8d_{3/2}$ levels are also drawn (85)

with 2 electrons and the $p_{3/2}$ subshell with 4 electrons. How large this effect can be is apparent from Fig. 7, where the energy eigenvalues of the $p_{1/2}$ and $p_{3/2}$ electrons are plotted as a function of $Z$ for the series of group-IV elements (65). This effect is of direct relevance for the chemical behavior of all elements in which these shells are the outer electron shells. This is, for example, the reason why in the group-IV elements the $+2$ valency becomes dominant for larger $Z$, as is already the case for lead, and why element 115 (eka-bismuth) is expected to have a monovalent state.

This spin-orbit stabilization also plays a dominant role in atomic lawrencium ($Z = 103$) with probably (66) a ground-state $7s^2 \, 7p_{1/2}^1$ instead of the expected $7s^2 \, 6d^1$, and in all the elements beyond 120.

As a summary of the calculations Fig. 8 shows the energy eigenvalues of all outer electrons for all elements between $Z = 100$ and 172 and in Fig. 9 the radii of the outermost electra wavefunctions for the elements 104 to 121 and 156 to 172 are shown.

In addition to the relativistic spin-orbit splitting, there are two more relativistic effects whose trends toward a chemical interpretation can be seen directly from the results of the calculations. The first is the so-called direct relativistic effect, which means the increase in the binding of the $s_{1/2}$ and $p_{1/2}$ levels relative to the nonrelativistic calculations. These $s$ and $p_{1/2}$ levels, even for large main quantum numbers, have wave functions that are nonzero in the vicinity of the nucleus, where the potential is large and the relativistic effects are increasingly strong with increasing $Z$. This effect explains why the $7s$ electrons are so strongly bound at the end of the $6d$ transition series where, instead of the expected increased full $6d$ shell stability, the $7s^2$ electrons remain bound at the elements 110 and 111,

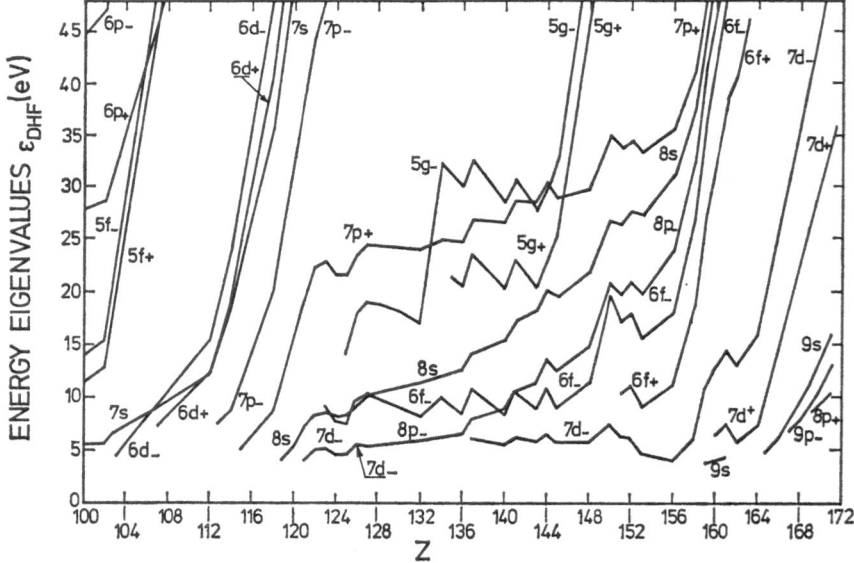

Fig. 8. Energy eigenvalues of the outer electrons for the elements 100 to 172 from DF calculations (35)

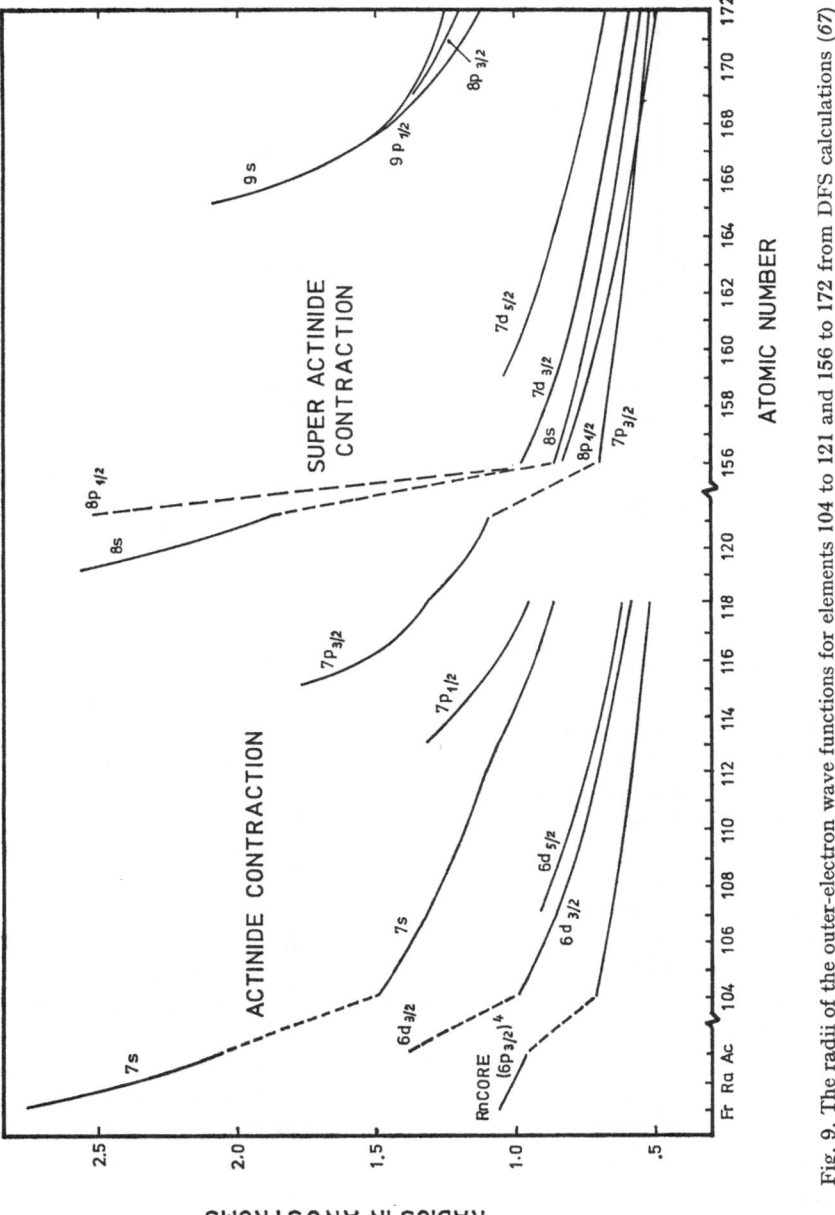

Fig. 9. The radii of the outer-electron wave functions for elements 104 to 121 and 156 to 172 from DFS calculations (67)

and possibly also at the ionized states of element 112, which would drastically change their chemical behavior relative to the extrapolated trends. This direct relativistic effect is also the reason why the trend of the decreased binding of the s electrons in the alkaline and alkaline earth metals stops at Fr and Ra, and the next metals of these groups are expected to have increased binding again. Parallel

105

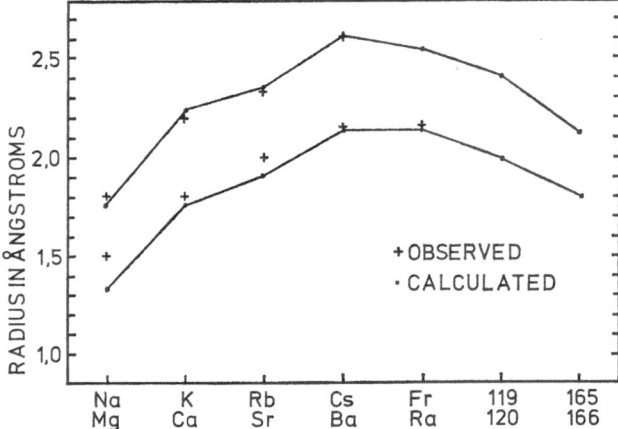

Fig. 10. Comparison of experimental and calculated atomic radii of the alkali and alkaline earth elements from DFS calculations (67)

to this result we show in Fig. 10 the calculated and experimental radii of these elements, taken from *Fricke et al.* (67), which shows excellent agreement. The conclusion from these calculations is that Cs is expected to be the largest atom in the entire periodic system, at least up to element 172.

The second relativistic effect to emerge from the results of the calculations is the so-called indirect relativistic effect. This effect describes the increased shielding and therefore decreased bonding of the electrons with large angular momentum, because the wave functions with small angular momentum are increased in binding and drawn into the atom because of the direct relativistic effect, thus shielding the other electrons more strongly. This effect is expected to occur, even from a chemical point of view, together with the large spin-orbit splitting effect at the end of the $7p$ elements at 118, where a noble gas element is actually located but a very reactive element with an easily obtainable $4^+$ state would be expected because the last four $7p_{3/2}$ electrons are so loosely bound.

Now, we have seen in this discussion that it is possible to calculate quite a number of physical quantities, and we understand from these calculations that several trends are a result of some physical influences, but there is still a long way to go to the prediction of chemical properties. Even the calculated quantities cannot be used as absolute but only as relative numbers, and the calculated trends have to be scaled to the experimental values in the known part of the periodic table. Therefore *all* the predictions of all quantities can only be done by a combination of the traditional method of continuing the trends of the interesting physical and chemical quantities of a chemical group into the unknown region of elements with a comparison of the trends shown in the calculation, if this is possible. But although the calculations sometimes do not give absolutely correct values, they are necessary as a guide to an unknown region, because the overall nature of the elements is being changed by relativistic effects.

**b) Empirical and Semi-empirical Methods.** A chemist who is trying to separate, for example, an unknown element from a sample needs to known more than just that it is similar to element xy. For an experienced chemist it is sometimes enough to know the anticipated location of the unknown element in the periodic table, because of his feeling for trends within the periodic table. To a certain degree, this sort of hunch cannot be expressed in scientific terms nor can it be calculated by ab-initio methods. A good review of the loose connection between electron configuration and chemical behavior in given by *Jørgensen (68)*.

The extrapolation of properties within either rows or groups of elements in the periodic table was and still is the best way of predicting the properties of unknown elements. There are quite a number of empirical and semi-empirical laws which have proved successful, ranging from the valence-bond theory (68), which must be used with care, via the Born-Haber cycle (69, 70) to *Jørgensens'* ingenious variations of this (71—73) to much more complicated extrapolations. For example, *David (74)* predicts thermodynamic quantities of the superheavy elements by plotting the log of the quantity versus the log of the atomic number. *Hoffmann* and *Bächmann (75, 76a, 76b)* used as plotting parameters for different properties $Zv/x$, $Z/x$ and $Zr^2/x$, with $Z$ the atomic number, $v$ the atomic volume, $x$ the electronegativity, and $r$ the covalent radius of the central atoms. Their results seem to be fairly good but the theoretical justification for these methods is lacking. They were trying to predict the properties of compounds of superheavy elements, especially the methyl, ethyl, hydrides and chlorides. [For their results we refer to the references (75) and (76)]. This of course is, at least from a theoretical point of view, another order of magnitude more complex. They confirmed their results by comparing them with values from independent extrapolations or from experiments in the known region of elements (76b, 76c). *Eichler (77)* did similar extrapolations to obtain information on the thermochromatographic separation of superheavy elements and compounds, which is another possible way of detecting and separating superheavy elements in small quantities.

The rel-HFS and rel-HF computer programs allow calculations of electronic energy levels, ionization potentials, and radii of atoms and ions from hydrogen into the superheavy region. In order to arrive at the oxidation states most likely to be exhibited by each superheavy element and also the relative stabilities of these various oxidation states, we need to be able to relate these properties to calculable electronic properties. The relationship between reduction potentials and the Born-Haber cycle has offered an effective approach to this problem (69, 70).

Electrode potentials are usually related to the standard $H^+/\frac{1}{2}H_2$ couple, whose potential is set equal to zero. We therefore consider the change in state for reduction of the aqueous metallic cation, $M^{n+}$(aq), to the metal, $M$(s):

$$M^{n+}(aq) + n\,\tfrac{1}{2}\,H_2(g) = M(s) + n\,H^+(aq) . \tag{1}$$

The change in Gibbs free energy is related to the reduction potential and to the enthalpy and entropy by the equation

$$- \Delta G = n\,E = - \Delta T + T\Delta S . \tag{2}$$

107

By considering all components in their standard states of unit activity or fugacity, we can obtain from these equations the standard electrode potential of the $M^{n+}/M(s)$ couple, as defined under the IUPAC Convention.

$\Delta S$ must be considered in each case, but so far, in most considerations pertinent to the superheavy region, it has either been chosen small or shown to be small (78). We shall therefore for simplicity consider only $H°$, which can be obtained through the Born-Haber cycle. First, the heat of sublimation, $S_M$, must be obtained through an extrapolation, preferably versus the row of the periodic tables, as has been done for several superheavy elements (e.g. for the discussion of elements 113 and 114). Secondly, the appropriate ionization energy, $I_n$, has to be calculated using ab-initio calculations. This value then has to be corrected, as discussed in the last paragraph. The difference between the calculated and experimental value in the known region of elements has to be extrapolated and then added to the calculated value.

The next part of the Born-Haber cycle is most conveniently taken to be the single-ion hydration energy, $H_{M^{n+}}$, although this quantity cannot be defined from a thermodynamic point of view. $H_{M^{n+}}$ can be obtained by simple extrapolation or by calculation, using various empirical modifications of the Born equation, depending on circumstances. For example, David (74) used a simple extrapolation to obtain 75 kcal (g atom)$^{-1}$ for $H_{113^+}$. Keller et al. (78) preferred to use the Born equation and obtained 72 kcal (g atom)$^{-1}$. These are then the quantities that make up $\Delta H°$, that is, the heat of sublimation of the metal $S_M$, the ionization energy $I_n$, and the single-ion hydration energy $H_{M^{n+}}$. Since we are not considering the entropy, we have for the change in state (1)

$$-\frac{1}{n}\Delta H° = E° = \frac{1}{n}\left[(I_n + S_M + H_{M^{n+}}) - n\left(\frac{1}{2}D_{H_2} + I_H + H_{H^+}\right)\right] \quad (3)$$

where $1/2\, D_{H_2}$, half the dissociated energy of the hydrogen molecule, is 2.26 eV; $I_H$, the ionization energy of the hydrogen atom, is 13.59 eV, and we accept the single-ion hydration energy of the proton as $-11.3$ eV, the value derived by Halliwell and Nyburg (79). This yields a value of 4.5 eV for the energy released when one gram equivalent of hydrogen ions is combined with electrons.

Equation (3) can be used to calculate the standard electrode potentials. Calculations based on the Born-Haber cycle to obtain the relative stabilities of oxidation states are known as "Oxidation State Diagrams". These diagrams have been found useful in clarifying inorganic chemistry (69), even though their accuracy is sometimes low.

Jørgensen (71—73) has formulated an ingenious approach for predicting the stabilities of oxidation states, based on Eq. (3). If we consider one-electron changes only, Eq. (3) can be rewritten

$$I_n - [E° + (\tfrac{1}{2}D_{H_2} + I_H + H_{H^+})] = -(H_{M^{n+}} + S_M) \quad (4)$$

or

$$I_n - (E° + 4.5) = -(H_{M^{n+}} + S_M). \quad (5)$$

*Jørgensen* calls $(H_M n + + S_M)$ the "hydration difference". $(E° + 4.5)$ is given the symbol $C_n$ and is called the "chemical ionization energy". *Jørgensen* finds empirically that he can set the "hydration difference" equal to $(2n - 1)k$ so that

$$I_n - C_n = (2\,n - 1)k \qquad (6)$$

where $k$ is a parameter that *Jørgensen* has determined for various different types of elements.

Since he wishes to predict states in aqueous solution, *Jørgensen* next notes that the range of oxidation states must be limited to those that neither oxidize nor reduce water. The oxidation reaction usually encountered in acid solution is

$$O_2 + 4\,H^+(aq) + 4\,e = 2\,H_2O\,. \qquad (7)$$

The standard electrode potential, $E°$, is 1.23 volts for this half-reaction.

Therefore, any couple of $E°$ larger than 1.23 volts can oxidize water, provided kinetic ("overvoltage") effects are absent. Since, under these conditions, $E°$ must be less than 1.23 volts, $C_n$ must be less than 5.7 eV.

*Jørgensen* expresses the condition that the ion will not oxidize water as follows:

$$I_n - (2\,n - 1)\,k = C_n < 5.7 \text{ eV}. \qquad (8)$$

On the other hand, from the thermodynamic point of view, any couple whose standard electrode (*i.e.* reduction) potential is negative will reduce water. *Jørgensen* writes this condition in the form appropriate to an oxidation reaction and obtains

$$I_{n+1} - (2\,n + 1)\,k = C_{n+1} > 4.5 \text{ eV}. \qquad (9)$$

On occasion, it may be worthwhile recalling that the potential of water oxidation reaction (*80*), Eq. (7), is a function of pH and oxygen pressure

$$E = 1.23 - 0.059\,\text{pH} + 0.0148 \log P_{O_2}\,. \qquad (10)$$

Similarly, the potential for water reduction reaction $[H^+(aq) + e = 1/2\,H_2(g)]$ is

$$E = -\,0.059\,\text{pH} - 0.0295 \log P_{H_2}\,.$$

In applying Jørgensen's approach, it should also be remembered that there are usually kinetic factors (so-called "overvoltage effects") which results in hydrogen and oxygen being evolved only at potentials beyond the range of the thermodynamic ones (*81*). There is also often the question of the effect of complex ion formation.

Valence-bond theory has proven valuable in understanding the chemistry of the known elements (*68*). As in the case of any inexact theory, it must be applied to cases where it can be expected to have validity. Making comparable calculations for the known elements will give us an idea of the validity of valence-bond theory for a specific superheavy element. We will not discuss this subject here in general, since details concerning the application of this method to the prediction of chemical properties may be found in the discussion of elements 111 and 115.

Another helpful scheme is the classification of ions with hard and soft Lewis acids by *Ahrland (82)* and *Pearson (83)*. While a specific definition or scale of softness is not universally accepted, the general principles are clear. They can be used to give somewhat more information for predictions of what compounds or complex ions might be expected for the superheavy elements.

Hard Lewis acids are found among the small, highly charged ions such as $Al^{+3}$ or $La^{+3}$, which have low-lying orbitals available for occupation — in general, these ions are not readily polarized. Soft Lewis acids tend to be large, easily polarized ions such as $Ag^{+1}$, and frequently the state of ionization is low. In addition, they contain unshared pairs of electrons such as $p$ or $d$ electrons in their valence shell. The hard Lewis bases are ions such as $F^-$ or $OH^-$, which have small ionic radii and are characterized by high electronegativity and low polarizability and are difficult to oxidize. In comparison to these ions, $I^-$ is more polarizable, acts as an electron donor and is therefore, a softer Lewis base.

Pearson's Hard-Soft-Acid-Base (HSAB) priciple is that hard acid-base combinations form readily and are generally ionic compounds. The other group of stable compounds and complex ions involves the interaction between soft acid and soft bases. For these, the bonding is primarily covalent with interpenetrating orbitals. The combinations hard acid with soft base, or vice versa, have little stability.

*Klopman (84)* defined the frontier orbitals for a base as the highest occupied orbitals of the donor atom or ion and the lowest unoccupied orbitals of the acid or acceptor ion. For elements with low atomic numbers, such orbitals are radially extended, *e.g.* the $3d$ orbitals. With increasing atomic number this feature is not guaranteed, and orbitals that are not tightly bound may become buried beneath other competitive orbitals. The essential idea is that the frontier orbitals must overlap significantly before covalent bonding can occur in the soft-soft inter-actions.

The difference in the chemistry of the light and heavy actinides may be rationalized in this way. The early members beyond thorium have unpaired $d$ and $f$ electrons available for forming covalent bonds and hence, for example, they readily form many complex ions and intermetallic compounds. Such ions are soft acids. Beyond americium, the $5f$ electrons are not competitive and the closed shell of six $5f_{5/2}$ electrons will not be readily available for bonding, so that only those $f$ electrons with $j = 7/2$ are available. These tend to become buried radially as the atomic number increases and hence their divalent ions become relatively hard Lewis acids. These considerations are especially helpful in the region of super-actinides because these elements do not have analogs in the known periodic table, where we have deeply buried but loosely bound $5g$ electrons.

# IV. Discussion of the Elements

The use of the continuation of the periodic table, the predicted electronic configurations, and the trends which become obvious from the calculations plus the semiempirical and empirical methods, allows us to offer some detailed predictions of the properties of the elements beyond lawrencium ($Z = 103$) $(85)$. Of course, these elements will first be produced at best on a "one atom at a time" basis, and they offer little hope of ultimate production in the macroscopic quantities that would be required to verify some of these predictions. However, many of the predicted specific macroscopic properties, as well as the more general properties predicted for the other elements, can still be useful in designing tracer experiments for the chemical identification of any of these elements that might be synthesized.

The most important property we need to know about an element is the stable oxidation states it can assume, because so many other chemical and physical properties depend upon the oxidation state. The second most important property to know concerns the relative stabilities of these oxidation states; that is to say, we need to know the standard electrode potential. As will be clear from the discussions in Section III.2b on this subject, the understanding of these two outstanding characteristics of the elements involves at least a knowledge of heats of sublimation, ionization potentials, ionic and atomic radii, and electronic energy levels. In this paragraph we try to focus on these properties, but we also summarize all the other properties so far predicted for the superheavy elements.

## 1. The 6d transition Elements $Z = 104$ to 112

Although elements 104, 105 (and very recently also 106) are known, they have not yet been much studied chemically. *Zvara* and coworkers $(86)$ believe they have shown element 104 to be tetravalent but there is no confirmation from other experimental groups on this. A first discussion of the chemical separation of element 105 is also given by *Zvara* $(87)$.

The electron configurations of the outer electrons of almost all the $d$-transition elements are given by the rule $(n-1)\, d^m n s^2$, where $n$ is the number of the period (or the principal quantum number) and $m$ goes from 1 to 10. This rule is not exactly valid in the 4th and 5th periods, where in some cases there is only one, or even no electron in the outer $s$ shell. This behavior is well understood; both the increasing binding energy of the $d$ electrons and the greater shielding of the $s$ electrons with higher $Z$, as well as the stabilizing effect of the half-filled and filled $d$ shell, lead to configurations where the number of $ns$ electrons is less than two $(88)$.

The main difference between the elements of the 4th and 5th periods and those of the 6th period is due to the fourteen $4f$ electrons, which are filled in between the occupation of the $6s^2$ and the $5d$ electrons. The filling of the $4f$ shell shields the less penetrating $5d$ orbitals more effectively than the $6s$ because the $d$ electrons are more fully barred from the inner parts of the atom by the centrifugal force $\frac{l(l+1)}{r^2}$. This means that the $d^m s^2$ configuration has increased stability in the 6th period, except at the high end where Pt and Au occur. This stabilizing effect of the $d^m s^2$ configuration becomes even larger in the next period, *i.e.* for the $6d$ elements, because the $7s$ electrons drop relatively deep into the atom. Thus they feel the very strong potential near the nucleus whereas the $6d$ electrons are shielded more strongly by the $5f$ electrons. Fig. 11 shows the calculated energy eigenvalues (*56, 85*) by a DFS method of the outer $s$ and $d$ electrons for the $d$ transition elements of the 6th, 7th and 8th periods. The trend is clear: the $s$ electrons are more strongly bound in the higher periods whereas the $d$ electrons are lowered and, in addition, the $d$ shell is split more and more into its two subshells, and the energy separation increases. Taken altogether, this explains why the electronic configuration of the neutral atoms of the transition elements in the 7th period will be given exactly by the rule $6d^m 7s^2$, as verified in the calculations shown in Table 2. The pairing energy that stabilizes filled and half-filled shells is no longer sufficient to break this general rule. Furthermore, pairing energy is as large as 1.1 eV per pair in the $3d$ elements and decreases for the higher periods. Cunningham (*88*) predicted this pairing energy to be less than 0.2 eV in the 7th period, so that its influence will be very small.

Fig. 12 shows the experimental first ionization energies of the transition elements for the $5d$ elements as well as the calculated values for the $5d$, $6d$, and $7d$ elements. This figure is located below Fig. 11 so that the trend of the ionization energy curves can be compared directly with the trend of the energy eigenvalues of the outer electrons. The ionization of the $5d$ electrons at the beginning of the $5d$ elements and the ionization of the $6s$ electrons at the end leads to the most stable configurations of the ion. The decrease in ionization energy between W and Re comes from the spin-orbit splitting. The $5d_{3/2}$ subshell, which holds only four electrons, is completed at W and the occupation of the less tightly bound $5d_{5/2}$ subshell begins with Re. The linear behavior thereafter is not continued for Pt and Au, where the $5d$ shell is closed sooner at the cost of the $6s$ electrons.

However, in the $6d$ elements the ionization of one $6d$ electron always leads to the most stable ion so that the ionization energy curve is nearly parallel to the energy eigenvalues of the most loosely bound $6d$ electron, as can be seen in Figs. 11 and 12. The calculated values are augmented by 0.2 eV, which equals the average difference between the experimental and calculated values for the 6th period. These values are expected to be correct to within $+0.4$ eV. (The values in Table 3 are taken for DF calculations.) But in ionic compounds first the $7s$ electrons, which are in the frontier orbital, have to be removed; their ionization energy is nearly 3 eV higher at the beginning and about 1 eV higher at the end of the $6d$ elements, which can also be seen in Fig. 12. Nevertheless, the relatively low ionization energy at the beginning of the $6d$ elements will be an indication that the maximal oxidation states for the $6d$ transition elements will be higher than, or at least equal to those in the $5d$ elements.

*Cunningham* (*88*) has given estimates of the sum of the first four ionization energies for elements 104 to 111. He uses extrapolation against the row of the

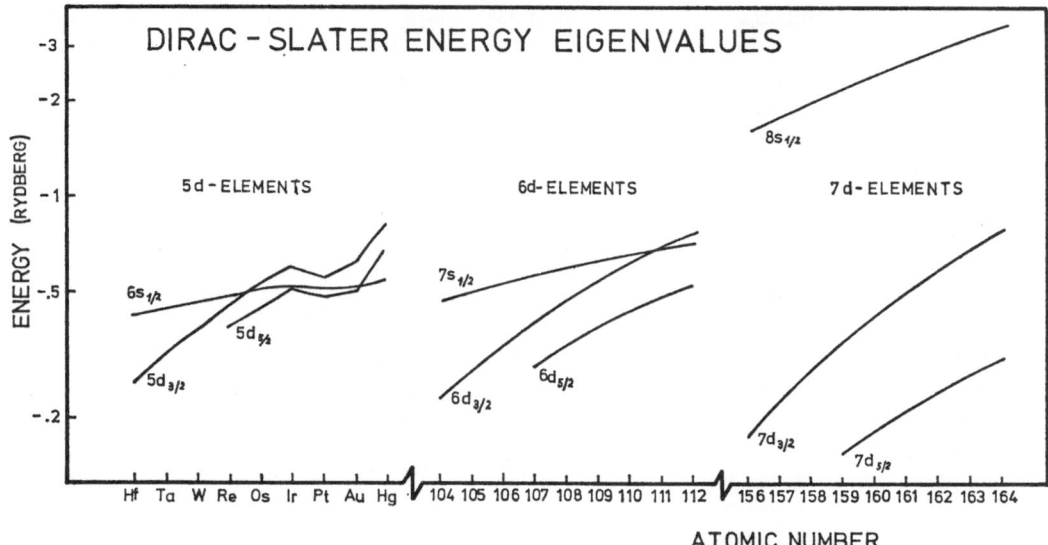

Fig. 11. Dirac-Slater (DFS) energy eigenvalues of the outer electrons for the 5d, 6d and 7d elements. This shows the strong relativistic increase of the binding of the last s shell and the increase in the splitting of the d subshells (85)

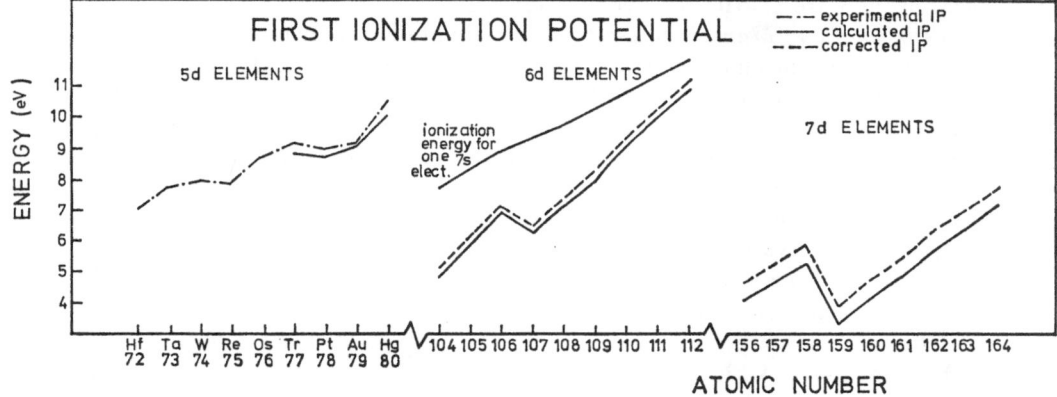

Fig 12. First ionization energies for the 5d, 6d and 7d elements. For the 6d elements the calculated removal energy is also given for one 7s electron, which is in the outermost shell (85)

periodic table to obtain the magnitude for 104 and then assumes that the increase along the 6d series as a function of Z will roughly parallel the 3d and 5d series. *Fricke* and *Waber* (85) and *Penneman* and *Mann* (89) have shown that Cunningham's values will be too low by about 10%, which leads to very high values at the end of the series, but also indicates that their chemical behavior is tending toward that of inactive noble metals with small maximal oxidation states.

113

*Penneman* and *Mann* (*89*) used Jørgensen's equations (8) and (9) with a $k$ of 4.5 eV to predict the most stable states of elements 104—110 in aqueous solution. Jørgensen's selection of $k$ applies to the hydrated cation (*72, 73*) and is not intended to account for the effects of complex ion formation or pH. The graph and tables of *Penneman* and *Mann* (*89*) yield the following most stable states in aqueous solution: 104(+4); 105(+5); 106(+4); (107+3); 108(+4); 109(+1); 110(0). In some cases, such as the +3 and +4 states of 108, the energies suggest they may be essentially as stable as the one listed.

*Cunningham* (*88*), using extrapolation, against the row of the periodic table, assumes that the pattern seen in oxidation states on going from the $3d$ to the $5d$ elements will continue into the $6d$. He obtains for the stable oxides: 104(+4); 105(+5); 106(+6); 107(+7); 108(+8); 109(+6); 110(+6). These results start deviating from those of *Penneman* and *Mann* at element 106. If solution conditions appropriate to the formation of oxyanions are established, oxidation states higher than those given by *Penneman* and *Mann* for 106, 107 and 108 would no doubt be attained, since the higher oxidation states of W, Re and Os are stabilized in this way. Such conditions would be closer to those assumed by *Cunningham*.

Radii for the $6d$ transition elements are given by *Fricke* and *Waber* (*85*). Their atomic and ionic radii were obtained using the principal maxima of the outermost electrons from DFS calculations. The metallic radii were obtained by complicated extrapolations of comparisons of the experimental trends as well as of the calculated trends of the radii of the outermost $s$ and $d$ electrons, taking into account empirically the fact that the outer electrons become more or less itinerant conduction electrons and are distributed over the crystal.

*Penneman* and *Mann* (*89*) also give an estimate of the metallic radii of the $6d$ elements, using for guidance the change in the calculated atomic radii of the $5d$ elements relative to the $6d$ elements as well as the experimental values. Their results are about 0.15 Å smaller than those of *Fricke* and *Waber*, involving a strong increase in density. Both estimates are listed in Table 3. One should bear in mind that both estimates are very rough extrapolations and can be used only as a first approximation. Estimates of the properties of the first four $6d$ elements were also given by *Häissinski* (*90*).

In the following we discuss each of the elements 104 to 112 separately. A list of their predicted properties is given in Table 3, the main references being (*85, 88—92*).

**Element 104.** (eka-hafnium) is predicted to resemble its homolog hafnium (element 72) in its chemical properties. It is expected to be predominantly tetrapositive, both in aqueous solution and in its solid compounds, although it should exhibit solid halides and perhaps aqueous ions of the +2 and +3 oxidation state as well.

One probably can predict some of the crystallographic properties, of the tetrapositive element 104 by extrapolation from those of its homologs zirconium and hafnium. The ionic radii of tetrapositive zirconium (0.74 Å) and hafnium (0.75 Å) suggest an ionic radius of about 0.78 Å for tetrapositive element 104, allowing for the smaller actinide rather than lanthanide contraction. Further one would expect the hydrolytic properties of element 104 and the solubilities of its compounds (such as the fluoride) to be similar to those of hafnium. The sum of the ionization potentials for the first four electrons should be less than that for hafnium, which suggests that it should be easier to oxidize element 104 to the $+4$ ionic state although, of course, the formation of the $+4$ state in covalent form will not require the complete removal of all four electrons.

**Element 105.** (eka-tantalum) should resemble tantalum and niobium, with the pentavalent state being the most important. It should exhibit several oxidation states, such as $+4$ and $+3$, in addition to the more stable $+5$ state. There should be an extensive range of complex ions, offering a rich chemistry.

**Element 106.** The chemical properties of element 106 (eka-tungsten) are predicted to be similar to those of tungsten, molybdenum and to some extent chromium, offering an even richer chemistry of complex ions than these elements. The hexafluoride should be quite volatile and the hexachloride, pentachloride and oxychloride should be moderately volatile. Penneman and Mann predict a $+4$ oxidation state in aqueous solution. Jørgensen's selection of $k$ is for the hydrated cation and is not intended to account for the effects of complex ion formation. However, since tungsten is stabilized in the oxidation state of $+6$ by the tungstate ion, an analogous situation may be expected for element 106.

**Elements 107 and 108.** Element 107 should be an eka-rhenium (with a volatile hexafluoride) and element 108 an eka-osmium, which suggests that the latter should have a volatile tetraoxide that would be useful in designing experiments for its chemical identification.

The differences in predictions of ionization states are large. The simple extrapolation of Cunningham suggests $+7$ and $+8$, whereas Penneman and Mann give $+3$ or $+2$ as the most stable state in aqueous solution. *Fricke* and *Waber* (67) predict $+5$ or $+6$. These inconsistent predictions may be taken as an indication that all of these oxidation states may exist, as is clear from the many possible oxidation states of the homologs Re and Os. Differences are expected in these elements, however, because of possible larger ligand field effects due to the greater spatial extension of the 6$d$ orbital charge cloud, as stated by *Cunningham* (88).

**Elements 109 and 110.** The same arguments are expected to apply to elements 109 (eka-iridium) and element 110 (eka-platinum) but in much clearer fashion. The ionization energies will be much increased and the metals are expected to have an even nobler character. If the higher oxidation states $+6$ and $+8$, as predicted by *Cunningham*, are stable volatile hexa- and possibly octafluorides will form, which may be useful for chemical separation purposes. As with elements 107 and 108, this is contradicted by *Penneman* and *Mann* (89), who arrive at $+1$ and 0 as the most stable oxidation states in aqueous solution, and *Fricke* and *Waber* (67), who give $+3$ or $+2$ as the dominant oxidation states. Again, nearly all these states are expected in the actual chemistry of these elements. The prediction of the possible oxidation state 0 means that it might well be that these elements at the end of the 6$d$ transition elements will remain as neutral atoms in the chemical separation processes, so that this method might fail. *Averill* and *Waber* (62) used the molecular orbital approach to estimate the stability of the hexafluoride of element 110. Their results indicate that this should be approximately as stable as the well-known platinum hexafluoride. Elements 109 and 110 should both have a strong tendency toward the formation of complex ions in their chemistry.

Table 3. Summary of predictions for elements 104 to 112

| Element | 104 | 105 | 106 | 107 | 108 | 109 | 110 | 111 | 112 |
|---|---|---|---|---|---|---|---|---|---|
| Chemical group | IVB | VB | VIB | VIIB | VIII | VIII | VIII | IB | IIB |
| Stable oxidation states [a] | 4, 3 | 5, 4, 3 | 6, 4 | 5, 7, 3 | 3, 4, 6, 8 | 3, 6, 1 | 4, 2, 6, 0 | 3 | 2 |
| First ionization energy (eV) calc. (DF)[b] | 5.1 | 6.6 | 7.6 | 6.9 | 7.8 | 8.7 | 9.6 | 10.5 | 11.4 |
| best expec. value | | | | 5.9 | 6.9 | 8.3 | 9.9 | 10.7 | 11.4 |
| Second IP (eV) (DFS) extr. | | | | 18.7 | 17.6 | 18.9 | 19.6 | 21.5 | 21.1 |
| Third IP (eV) (DFS) extr. | | | | 27.9 | 29.2 | 30.1 | 31.4 | 31.9 | 32.8 |
| Fourth IP (eV) (DFS) extr. | | | | 37 | 38 | 40 | 41 | 42 | 44 |
| Fifth IP (eV) (DFS) extr. | | | | 49 | 52 | 51 | 53 | 55 | 57 |
| Atomic radius (Å)[d] | 1.49 | 1.42 | 1.36 | 1.31 | 1.26 | 1.22 | 1.18 | 1.14 / 1.2 | 1.10 |
| Ionic radius (Å)[d] | (+4) 0.71 / 0.78 | (+5) 0.68 | (+4) 0.86 | (+5) 0.83 | (+4) 0.80 | (+3) 0.83 | (+2) 0.80 | (+3) 0.76 | |
| Metallic radius (Å)[e] / [c] | 1.66 / 1.50 | 1.53 / 1.39 | 1.47 / 1.32 | 1.45 / 1.28 | 1.43 / 1.26 | 1.44 / 1.28 | 1.46 / 1.32 | 1.52 / 1.38 | 1.60 / 1.47 |
| Density (g/cm$^3$)[e] / [c] | 17 / 23.2 | 21.6 / 29.3 | 23.2 / 35.0 | 27.2 / 37.1 | 28.6 / 40.7 | 28.2 / 37.4 | 27.4 / 34.8 | 24.4 / 28.7 | 16.8 / 23.7 |
| Heat of sublimation (kcal/mol) | | | | 220 | 201 | 180 | 210 | 125 | small |
| Standard electrode potential ($V$)[f] | 0 → +4 / < 1.7 | | | | | | | | |
| Melting point (°C)[f] | 2100 | | | | | | | | |
| Boiling point (°C)[f] | 5500 | | | | | | | | |

Electron affinity (eV) ~1.6

a) The underlined values are the oxidation states in aqueous solution as given in Ref. (89). b) Ref. (35). c) Unpublished. d) Ref. (67).
e) Ref. (56). f) Ref. (26) in Ref. (85). g) Ref. (93).

**Element 111.** (eka-gold) has been studied in great detail by *Keller, Nestor, Carlson* and *Fricke (93)*. They used valence-bond theory to predict the most stable oxidation state $+3$ of this element. In Au($+3$) compounds the ligands form a square-planar arrangement around the central Au ion, indicating $dsp^2$ hybridization of its orbitals. The $5d$, $6s$, and $6p$ electrons are used for bonding and another pair of electrons is accepted from an anion to form complexes such as $AuCl_4^-$. The promotion energy from the $6d^{10}6s^1$ ground state to the $5d^96s^16p^1$ is about 5.9 eV compared with 6.2 eV for the promotion from the $6d^97s^2$ ground state in element 111 to the analogous hybridized configuration. Because the atomic radius of 111 will be 1.2 Å, whereas that of gold is 1.35 Å, element 111 should form at least as strong bonds as Au since the hybridization orbitals are not diffused over as large a volume. Also, the heats of sublimation of 111 and Au should be very similar because the breaking of the full $d$ shell will be nearly compensated for by the filled $s$ shell. The smaller atomic radius of 111 and its higher ionization potential suggest further that the heat of sublimation may be slightly higher rather than slightly lower.

These energy-expending processes, the heat of sublimation and the promotion to the $+3$ state, will be more than compensated for by the energy released in bond formation because the smaller radius of 111 will allow larger orbital overlap with ligand orbitals. *Keller et al.* therefore expect 111 to be stable in the oxidation state $+3$ and to have a chemistry similar to Au($+3$). The $+1$ state of element 111 will be very unstable and, if it exists at all, it will be in complexes with strongly polarizable ligands such as cyanide. 111($+2$) is not expected to exist. An interesting feature will be the possible stability of the $111^-$ ion, analogous to the auride ion. The expected electron affinity of element 111 lies between that of Au, which forms the negative auride ion in chemical compounds like CsAu and RbAu, and Cu and Ag which do not.

**Element 112.** Similar detailed considerations concerning the chemical properties have not yet been made *(85, 91, 94)* for element 112 (eka-mercury). More qualitative conjectures suggest that the most stable oxidation states will be the $(+1)$ and $(+2)$ states, but that higher oxidation states will probably be important in aqueous solution and in compounds. In macroscopic quantities it should be a distinctly noble metal, but because both the $6d$ and $7s$ outer electron shells are filled and their ionization energies are higher than in all the homologs, one may argue that the interatomic attraction in the metallic state will be small, possibly even leading to high volatility as in the noble gases. Element 112 should have an extensive complex ion chemistry, like all the elements in the second half of the $6d$ transition series. A general feature of these elements is the expected marked tendency to be "soft" acceptor ions. In Fig. 13, in which the logarithm of the stability constant for the halide compounds is plotted for the subgroup-IIB elements, a marked increasing softness for the higher elements can be seen. *Kratz* and co-workers *(94)* concluded from this figure that $112^{+2}$ should form very stable iodine and bromine complex ions, which would be extremely useful for the chemical separation. However it is also possible that this element might show a very low chemical activity and simply remain in its neutral atom configuration.

In Table 3 the predicted properties of the $6d$-transition elements are briefly summarized.

117

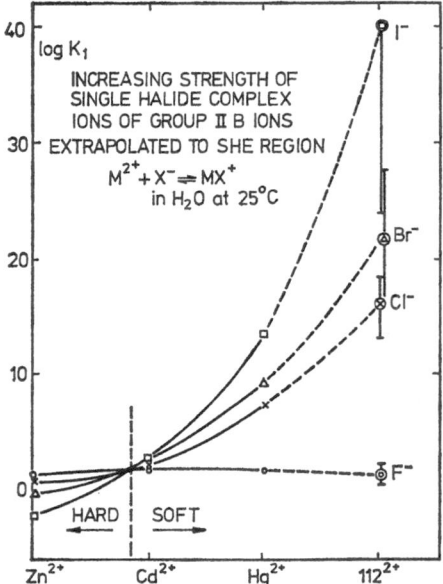

Fig. 13. Logarithm of the stability constant for the halide compounds of the group-IIB elements (94)

## 2. The 7p and 8s Elements Z = 113 to 120

The lifetimes of the elements near $Z = 114$ are expected to be years or even longer. Hence a knowledge of their chemical behavior is most important, since it might even be possible to find small amounts of those elements on earth, although the experimental sensitivity of these searches has reached $10^{-14}$g/g with no clear evidence of the existence of superheavy elements (32).

As can be seen from the Hartree Fock calculations in Table 2, from elements 113 to 118 the 7p electrons will be filled in, elements 119 and 120 following with the 8s electrons. The energy eigenvalues from DFS calculations (85) of the outer electrons of these elements are compared with the analogous values for the 6p elements Tl to Ra in Fig. 14. The great similarity in the occupation pattern of the valence electrons is evident. However, there are two significant differences. The analogous s electrons are bound more tightly for higher $Z$; this binding results from the large direct relativistic effects, as discussed in Section III.2a. The spin-orbit splitting of the 7p shell has increased compared to the 6p shell. A complete subshell with quantum numbers $|nlj\rangle$ is spherically symmetric, just as the complete nonrelativistic $|nl\rangle$ shells are, which means that there will be a change in the angular distribution of the electrons in the relativistic treatment, wich could lead to differences in chemical behavior for incomplete $|nlj\rangle$ subshells of very heavy elements. The first ionization potentials of the $p$ elements can be seen in Fig. 15, where the experimental ionization energies for the 3rd, 5th and 6th periods are shown together with the calculated values for the 6th, 7th and

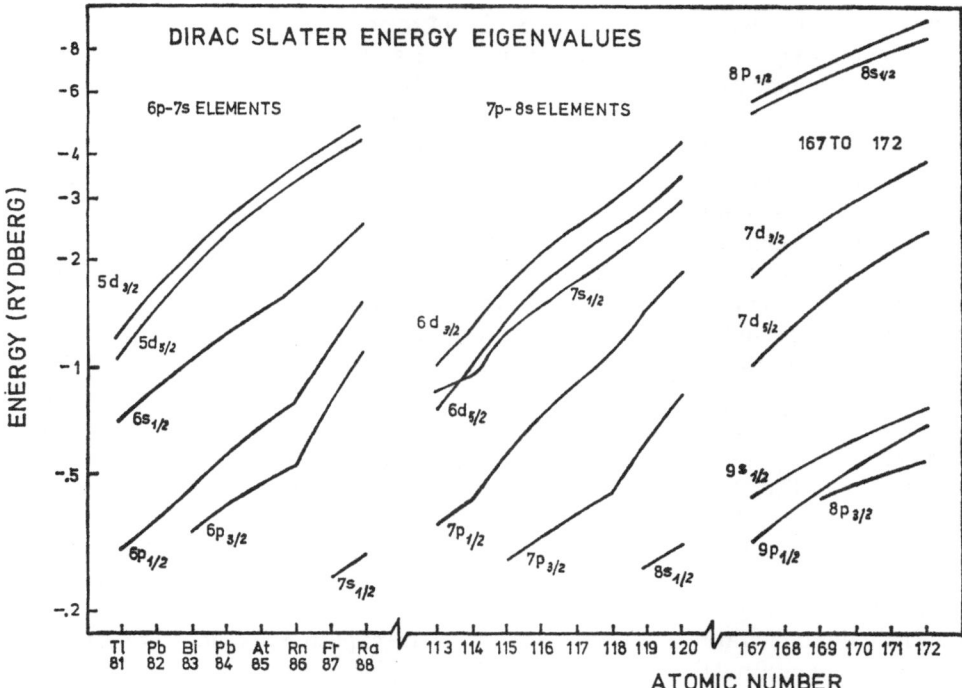

Fig. 14. Dirac-Slater (DFS) energy eigenvalues of the outer electrons for the 6*p*—7*s* and 7*p*—8*s* elements as well as for elements 167 to 172. This figure demonstrates the strong increase of the splitting of the *p* subshells. For elements 167 to 172, the 9*p*$_{1/2}$ and the 8*p*$_{3/2}$ subshells are energetically so close to each other that they form a *p* shell which is comparable to the 3*p* shell (*85*)

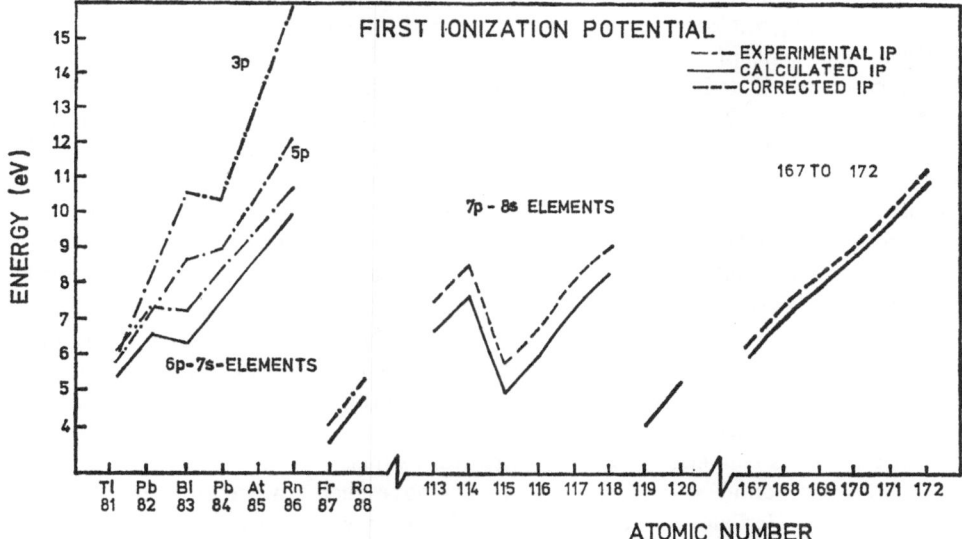

Fig. 15. First ionization energies for the 6*p*—7*s* and 7*p*—8*s* elements as well as elements 167 to 172. For comparison, first ionization energies for the 3*p* and 5*p* elements are also given (*85*)

9th periods. This figure is located below Fig. 14 to assist comparison of the trends of the energy eigenvalues and ionization energies. The ionization energy in the early periods increases linearly with only one break between configurations $p^3$ and $p^4$. This behavior is caused by the half-filled shell where three electrons occupy the three possible $p$ orbitals with parallel spin. The pairing energy is negative so that the ionization energy is smaller for the 4th electron. This effect of the Russell-Saunders coupling is very significant in the first periods but becomes smaller for the higher periods, because the LS coupling changes to intermediate coupling and from there to $j$—$j$ or spin-orbit coupling. In the 6th period, the break between $p^3$ and $p^4$ has vanished but instead a break occurs between $p^2$ and $p^3$; this trend is expected, since the spin-orbit coupling becomes more important than pairing energy in the $p_{1/2}$ and $p_{3/2}$ subshells. This break becomes even larger for elements 113 to 118. Thus, the ionization energy of element 114 is nearly as large as the same value for the "noble gas" that occurs at $Z = 118$. The experimental and calculated curves for the 6th period are almost parallel with a difference of about 0.8 eV. Therefore, the first ionization energies of elements 113 to 118 have been enlarged by 0.8 eV to get more realistic values (56). The estimated error is expected to be about $+0.4$ eV. Of course, it is even better to extrapolate these differences for every chemical group separately, as is done in detailed discussion of the elements (78).

The oxidation states of elements 113 to 118 are expected to follow the systematic trends for group- III to VIII elements, that is, elements with higher $Z$ prefer the lower oxidation states. In the first part of the $p$ elements this is the expression of the fact that the s electrons with their greater binding energy lose their chemical activity as $Z$ increases, so that the $p$ electrons will be the only valence electrons available; in the second part the same holds true for the $7p_{1/2}$ electrons, which form a spherical closed shell (at least in $j$—$j$ coupling), so that only the $7p_{3/2}$ electrons will remain chemically active.

In predicting the atomic and ionic radii of these elements, one gets into difficulties. The concept of atomic radii of *Slater* (63), who says that the atomic radius can be well defined as being the radius of the principal maximum of the outermost electron shell, *i.e.* of the frontier orbital, works quite well in most parts of the periodic system. However, at the end of the $d$ transition elements and for most of the $p$ elements, this definition apparently yields inaccurate results. The reason for this behavior is not quite clear but it may be connected with the type of bonding, *i.e.* the large hybridization effects in the compounds of these elements. Therefore a continuation of the trends of the behavior of the metallic or ionic radii, as done by *Grosse* (95), *Keller et al.* (78), and *Cunningham* (96), is expected to give more accurate results. *Fricke* and *Waber* (56) predicted the metallic radii of these elements on the basis of a comparison of the computed total electron density at large radii of known elements. The metallic radius of the superheavy elements was assumed to be where the density was equal to the density of the known analogous elements at its experimental radius.

Several compounds of elements 112 to 117 have been studied theoretically in great detail by *Hoffmann* and *Bächmann* (75, 76) as well as by *Eichler* (77), because the gaseous hydrids, chloride and a few other compounds, may be used for quick chemical separation by the method of gas chromatography. We do not list their straight extrapolations in the tables because it would be necessary to give for every compound the values of boiling point, heat of sublimation, heat of vaporization, enthalpy, heat of formation, dissociation energy, ionization potential, etc. For these values, we therefore refer to the literature.

**Element 113.** Detailed predictions of element 113 (eka thallium) have been given by *Keller et al.* (*78*). In this chemical group III A the main oxidation state is 3; only Tl has also a monovalent state which is associated with the ionization of the single $6p_{1/2}$ electron and the relatively increased stability of the 6s electrons. From this behavior, the high ionization energy of the $7p_{1/2}$ electron, and the even more increased stability of the 7s electrons, the principal oxidation state of element 113 is expected to be $+1$.

The straightforward extrapolations of *Keller et al.* of the physical and chemical properties of elements 113 and 114 are mainly based on the fact of the simple outer electronic structure of $7p_{1/2}$ and $7p_{1/2}^{2}$, with their large energetic and radial separation to the 7s and 6d electrons. As an example of these predictions, which are listed in Table 5, Fig. 16 shows the very suggestive extrapolations of the heat of sublimation of elements 113 and 114, and Fig. 17 the extrapolation of the melting point of element 113. The second important quantity for the derivation of the standard electrode potential through the Born Haber cycle is the ionization

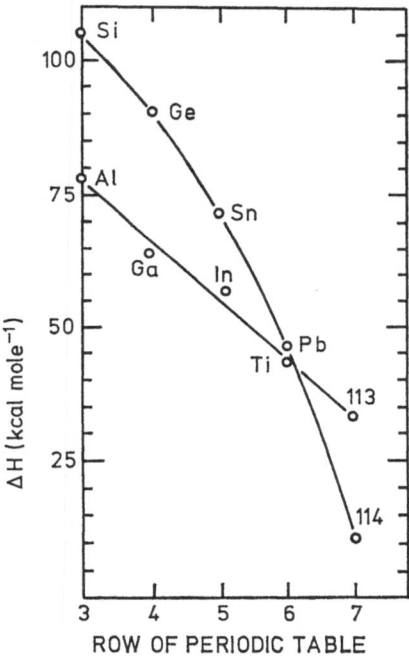

Fig. 16. The evaluation of the heat of vaporization for elements 113 and 114 is shown as an example of the extrapolation of chemical and physical properties, as performed by *Keller et al.* (*78*)

121

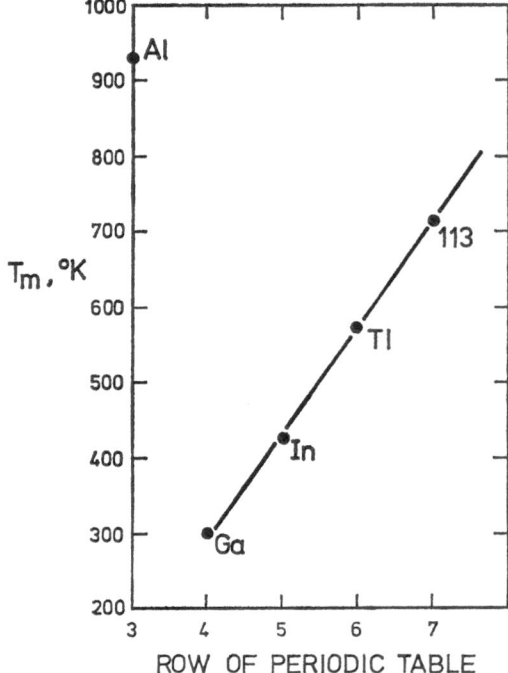

Fig. 17. Extrapolation of the melting point of element 113 (78)

potential. In Table 4 we show the method used to derive this quantity by means of both calculation and extrapolation. The value in parentheses is expected to be the best value. The third quantity needed is the single-ion hydration energy, $H_M n +$. *David* (74) obtained 75 kcal (g atom)$^{-1}$ for $H_{113} +$ whereas Keller (78) preferred to use the Born equation and obtained 72 kcal (g atom)$^{-1}$. These values lead to a

Table 4. Example of the extrapolation of calculated and experimental ionization energies to the best values for element 113. The calculated values are taken from DFS calculations (78)

| Element | First Ionization Potential (eV) (Group III A) | | |
| --- | --- | --- | --- |
| | calculated | experimental | $\Delta$ |
| Al | 4.89 | 5.98 | 1.09 |
| Ga | 4.99 | 6.00 | 1.01 |
| In | 4.87 | 5.78 | 0.91 |
| Tl | 5.24 | 6.11 | 0.87 |
| 113 | 6.53 | (7.36) | (0.83) |

standard electron potential of $+0.6$ volts for element 113, which indicates that element 113 will be somewhat less active than Tl.

To summarize the chemistry which can actually be expected for element 113, one may say that it is in general expected to fall between that of $Tl^+$ and $Ag^+$. $113^+$ is expected to bind anions more readily than $Tl^+$ so that $(113)Cl$ will be rather soluble in excess HCl whereas the solubility of TlCl is essentially unchanged. Similarly, $(113)Cl$ is expected to be soluble in ammonia water in contrast to the behavior of TlCl. The behavior of the $113^+$ ion should tend toward $Ag^+$ in these respects. Also, although $Tl(OH)$ is soluble and a strong base, the $113^+$ ion should form a slightly soluble oxide that is soluble in aqueous ammonia.

**Element 114.** (eka-lead) is of special interest because of its expected double-magic nucleus with a possible long lifetime. It belongs to the group-IV elements (78) where the stability of the oxidation state $+4$ decreases and $+2$ increases with increasing $Z$. The tetravalency in this chemical group IV is connected with a $sp^3$ hybridization and strong covalent bonding. The energetic difference between the $ns$ and the average of the $np$ electrons increases for the higher periods, as can be seen in Fig. 7, as the result of the relativistic effects. Also, the spin of the $p$ electrons couples strongly to their own angular momentum. Therefore, a $sp^3$ hybridization is no longer easily possible for 114, which means that the predominant oxidation state of 114 will be divalent. However, since the outermost $s$ and $d$ electrons in 114 have approximately equal energies, some form of $sd$ hybrid would be possible. Thus, one cannot exclude the possibility that a volatile hexafluoride might form. Because the $7p_{1/2}^2$ shell, in 114 approximates to a closed shell analogous to an $s^2$ closed shell such as is found in mercury, it is worthwhile comparing what one finds on going from 113 to 114 with what is known about going from Au (with a $s^1$ configuration) to Hg (with a $s^2$ configuration) in addition to the chemical group extrapolations. The results are given in Table 5. The standard electrode potential derived from these values by *Keller et al.* for element 114 is $+0.9$ volt, so that it is expected to be more noble than lead. If the error of $\pm 1.0$ volt is negative, the value would be about the same as lead. *Jørgensen* and *Haissinsky* (97) suggest that 114 may even be alkaline earth-like in its chemistry, which seems to be extremely far away from *Keller's* (78) results.

Summarizing, one can say due to relativistic effects the $+2$ valency of element 114 is strongly favored and is expected to resemble $Pb^{+2}$ chemistry, with a still greater tendency to form complexes in solution. In excess halogen acid, complexes of the type $114\,X_n^{(n-2)}$ should be stable. A complex analogous to the plumbite ion is expected. The sulfate and sulfide should be extremely insoluble, but the acetate and nitrate soluble. The latter may show extensive hydrolysis. *Hofmann* (98) discusses the probable presence of element 114 in nature as a result of its chemical behavior.

**Element 115.** A most intriguing situation is offered by element 115 (eka-bismuth) because a $7p_{3/2}$ electron is added here outside the $7p_{1/2}^2$ closed shell. As pointed out by *Fricke* and *Waber* (85), the $7p_{3/2}$ binding energy is much less than that of the $7p_{1/2}$ electrons. Consequently, they predict that element 115 will have $+1$ as its normal oxidation state. *Keller, Fricke* and *Nestor* (99) have recently obtained preliminary results indicating that the $+1$ state will act like $Tl^+$. This prediction is based on the values they have obtained for the ionization energy,

Table 5. Summary of predictions for elements 113 to 120

| Element | 113[e] | 114[e] | 115[f] | 116 | 117[g] | 118[g] | 119[g] | 120[g] |
|---|---|---|---|---|---|---|---|---|
| Chemical group | IIIA | IVA | VA | VIA | halogen | noble gas | alkali | alkaline earth |
| Stable oxidation states | 1,3 | 2 | 1,3 | 2,4 | 3,1,5,−1 | 4,2,6 | 1, others | 2, others |
| Ionization potentials (eV) | | | | | | | | |
| I best value | 7.4 | 8.5 | 5.5 | 7.5 | 7.7 | 8 7 | 4.8 | 6.0 |
| I DF calculation[a] | 8.0 | 8.9 | 5.5 | 6.6 | | 16.2 | 17.6 | |
| II (DFS) | 23.2 | 16.6 | 18.2 | 13.8 | | | | |
| III (DFS) | 33.2 | 34.9 | 27.5 | 29.5 | | | | |
| IV (DFS) | 45.1 | 45.6 | 48.5 | 39.5 | | | | |
| V (DFS) | 59 | 60.6 | 59.3 | 63 | | | | |
| Standard electrode potential (eV) | (0 → 1) +0.6 | (0 → 2) +0.9 | (0 → 1) +1.5 | | (−1 → 0) +0.25 −0.5 | | (0 → 1) 2.9 | (0 → 2) 2.9 |
| Atomic radius (Å) | | | 2.0 | | | | 2.4[h] | 2.0[h] |
| Ionic radius (Å) | (+1) 1.4 | (+2) 1.2 | (+1) 1.5 (+3) 1.0 | | | | (+1) 1.8 | (+2) 1.6 |
| Metallic radius (Å)[b,c] | 1.7 1.62 | 1.8 1.68 | 1.87 1.67 | 1.83 1.69 | | | 2.4 | 2.0 |
| Density (g/cm$^3$) | 16 | 14 | 13.5 | 12.9 | | | 3[k] | 7[k] |
| Melting point (°C) | 430 | 67 | 400 | | 350—550 | −15 | 0—30 | 680 |
| Boiling point (°C) | 1130 | 147[e] ~1000[l] 2840[m] | | | 610 | −10 | 630 | 1700 |
| Heat of vaporization (kcal/g-atom) | 31 | 9 | 33 | 10 | | | | |
| Heat of sublimation (kcal/g-atom) | 34 | 10 | 34 | 47 | | | | |
| Standard enthalpie (kcal/g-atom)[d] | 26±2 | 17±4 | 36±3 | 20±4 | 19±2 | | 10 | 33 |
| Entropy (cal deg$^{-1}$ (g-atom)$^{-1}$) | 17 | 20 | 16 | | | | | |

a) Ref. (35).  b) Ref. (56).  c) Ref. (89).  d) Ref. (77).  e) Ref. (78).  f) Ref. (99).
g) The properties below the ionization potentials are from B. B. Cunningham to be found in Ref. (5).
h) Ref. (85).  k) Ref. (91).  l) Ref. (98).  m) Ref. (108).

ionic radius, and polarizability, which show that 115⁺ is much more similar to Tl⁺ than to Bi⁺. An oxidation potential of 1.5 volts is predicted, indicating that the 115 metal is quite reactive.

The chemistry of the 115⁺ ion can be summarized as follows: the complexing ability of 115⁺ can be expected to be low with such anions as the halides, cyanide and ammonia. Hydrolysis should occur readily for 115 in the oxidation state of 1, and the hydroxide, carbonate, oxalate and fluoride should be soluble. The sulfide should be insoluble and the chloride, bromide, iodine, and thiocyanide only slightly soluble. For example, excess HCl will not appreciably affect the solubility of (115)Cl.

*Smith* and *Davis (100)* have recently discussed Bi⁺ chemistry in the hope that this may give some insights into 115⁺ chemistry. *Keller, Fricke* and *Nestor (99)* have also obtained a preliminary estimate of the stability of 115(+3) by analogy to Tl(+3). In their treatment they regard the $7p_{1/2}^2 7p_{3/2}$ valence state of 115(+3) to be analogous to the $6p6s^2$ valence state of Tl(+3). Consideration of the promotion energy and heat of sublimation of 115 relative to those of Tl, plus the expected overlap of its orbitals with those of ligands, lead to the conclusion that the oxidation state of +3 will be quite important in 115 chemistry besides the +1 state. But, as in the arguments used for the +1, state the 115⁺³ ion is expected to be most like Bi⁺³. The trichloride, tribromide, and triiodide of 115⁺³ will probably be soluble, and they may show a tendency to hydrolyze to form salts analogous to BiOCl and BiOBr. The trifluoride should be insoluble like BiF₃ as well as (115)S₃. The sulfate and nitrate will be soluble in the appropriate acids, and the phosphate will be insoluble.

It is not yet possible to predict the relative stabilities of the +1 and +3 states. In fact, their relative stabilities may well depend strongly on the state of complexation or hydrolytic conditions. On the other hand, element 115 will not show the group oxidation state of +5. The other properties of the element may be found in Table 5.

Because element 114 behaves to some extent like a closed-shell atom, and element 115 as if a new group of elements had been started, we give in Fig. 18 the results of *Eichler (77)* from the extrapolation of the standard enthalpy of elements 112 to 117. The relatively small value for 114 and the relatively large value for 115 may be taken as another indication that the interpretation given here, arising from the atomic calculations with the relativistic effects, will be important as regards many physical quantities as well as in the chemical interpretation of these elements.

**Element 116.** Not much work has been done on element 116 (eka-polonium) beside the normal extrapolations and calculations of the ground state and ionized configurations *(85)*. The values obtained are listed in Table 5. The chemical properties of element 116 should be determined by extrapolation from polonium; thus it should be most stable in the +2 state with a less stable +4 state.

**Element 117.** (eka-astatine) is expected to have little similarity to what one usually calls a halogen, mainly because its electron affinity will be very small. *Cunningham (96)* predicted its value as 2.6 eV, whereas the calculations of *Waber, Cromer* and *Liberman (54)* gave a value of only 1.8 eV. As a result of this small electron affinity, and from extrapolations of the chemical properties of the

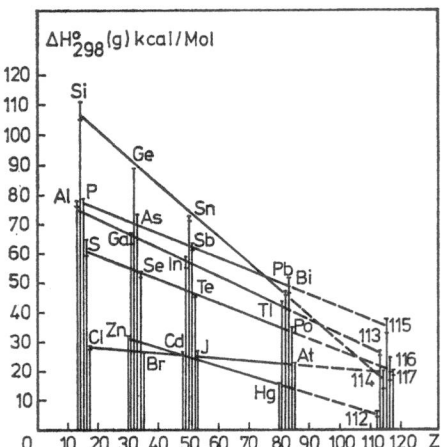

Fig. 18. Extrapolation of the standard enthalpy for elements 112 to 117, as given by *Eichler* (77)

lighter halogen homologs, all authors agree that the $+3$ oxidation state should be at least as important as the $-1$ state, and possibly more so. To take an example, element 117 might resemble Au($+3$) in its ion-exchange behavior with halide media. *Cunningham* (96) describes the solid element 117 as having a semi-metallic appearence.

**Element 118.** *A. V. Grosse* wrote a prophetic article (95) in 1965 before the nuclear theorists began to publish their findings concerning the island of stability. In this paper he gave detailed predictions of the physical and chemical properties of element 118 (eka-radon), the next rare gas. He pointed out that eka-radon would be the most electropositive of the rare gases. In addition to the oxides and fluorides shown by Kr and Xe, he predicted that 118 would be likely to form a noble gas-chlorine bond. These very first extrapolations into the region of super-heavy elements have been fully confirmed by the calculations, because the first ionization potentials turn out to be much lower than in all the other noble gases.

Independently *Grosse* (95) and *Cunningham* (96) found that the expected boiling point of liquid element 118 is about $-15\ ^\circ$C, so that it will be nearly a "noble fluid". Because of its large atomic number it will, of course, be much denser than all the other noble gases. But, in general, the chemical behavior of element 118 will be more like that of a normal element, with many possible oxidation states like $+2$ and $+4$; $+6$ will be less important because of the strong binding of the $p_{1/2}$ electrons. It will continue the trend towards chemical reactivity first observed in xenon.

**Elements 119 and 120.** In the two elements 119 (eka-francium) and element 120 (eka-radium) the $8s$ electrons will be bound very tightly and therefore these two elements are expected to be chemically very similar to Cs and Ba or Fr and Ra. Fig. 14 shows the energy eigenvalues of the outer electrons and Fig. 15 the ionization potential for elements 119 and 120 drawn in comparison to Fr and Ra. The main oxidation state of 119 and 120 will be 1 and 2, as is normal for alkali and alkaline earth metals. Their ionization potential will be about 0.5 eV higher than in the elements Fr and Ra, mainly because the $s$ electrons penetrate deep into the atom and feel the very strong potential near the nucleus. Therefore, their atomic radius according to the simple definition of *Slater* (*63*) is expected to be 2.4 and 2.0 Å, very similar to the values for Rb and Sr. This decreasing trend in the radius can best be seen in Fig. 10. The same holds true for the ionization energy displayed in Fig. 15. Therefore the early predictions of *Cunningham* (*96*), who extrapolated the trends of the known alkali and alkaline earth elements, have had to be corrected so that the chemistry of elements 119 and 120 will be closer to Rb and Sr than to Fr and Ra in the $+1$ or $+2$ oxidation states, respectively. On the other hand, the ions will have larger radii than $Rb^+$ and $Sr^{2+}$ because of the larger extension of the filled $7p$ shell in comparison to the lower $p$ shells, so that hydration will be more important and crystal energies will be different. Another important point is that higher oxidation states may be reached (*85*) in the presence of strong oxidizing agents because the ionization energy of the outer $7p_{3/2}$ electrons is only of the order of 10 eV. *Penneman* and *Mann* (*89*) came to the same conclusion; thus, an oxidation state of $+3$ and $+4$ should be considered.

Table 5 gives the chemical and physical properties of elements 119 and 120.

## 3. The $5g$ and $6f$ Elements $Z = 121$ to 154

The next elements of the periodic table, starting with element 121; belong to a very long, unprecedented transition series which is characterized by the filling of not only the $6f$ but also the $5g$ electrons. *Seaborg* (*5*) called these elements *Superactinides*.

Unfortunately, it is expected that the chemistry of these elements will not be able to be studied because the theoretical investigations of nuclear stability predict that these elements will be unstable and have very short lifetimes. Before this was known, a large number of theoretical calculations of the ground-state electronic configurations were made in this region because the proton number $Z = 126$ was long expected to be the center of the first island of stability. Now this is considered unlikely. Nevertheless, the chemistry of these elements would be very interesting.

In the lanthanoides and actinoides, the competition between the outer $d$ and inner $f$ electrons determines the ground-state electron configuration as well as the chemistry of these elements. Here at the beginning of the superactinides, not merely two but four electron shells, namely the $8p_{1/2}$, $7d_{3/2}$, $6f_{5/2}$ and $5g_{7/2}$, are expected to compete nearly simultaneously in the atom, and these open shells together with the $8s$ electrons determine the chemistry. The results of the ground-state calculations of these elements can be found in Table 2.

Three most interesting things occur in elements 121 to 154. First, the $8p_{1/2}$ electrons are filled, beginning with element 121, and at least one of these electrons remains in all following elements. This is clearly a direct relativistic effect, which is most effective for all $j = 1/2$ levels. Second, during the filling of all the superactinide elements some other electrons besides the $5g$ and $6f$ electrons always remain in the ground-state configuration in contrast to the analogous lanthanoides and actinoides, where at the beginning some $d$ electron states are occupied but are

127

removed during the filling of the $f$ shell. Third, the effective binding of an electron with large values of the angular momentum is accompanied by the radial collapse of the orbital, the centrifugal term $l(l+1)/r^2$ keeping it extended. For example, the effective radius of the $5g$ electrons changes from 25 Bohr units in element 120 in the excited configuration $8s^15g^1$ to 0.8 for element 121 in the configuration $8s^17d^15g^1$, according to *Griffin et al.* (*101*), as shown in Fig. 19. This demonstrates that large changes occur in this case, although the change in the effective potential is relatively small. The best calculations which are from *Mann* (*35, 50*) show that this collapse actually occurs as late as element 125 as a consequence of the indirect relativistic effect.

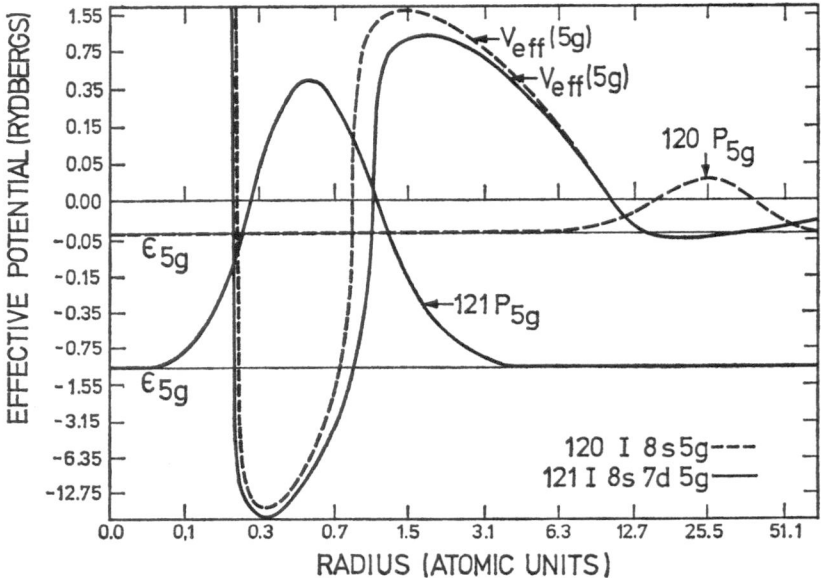

Fig. 19. Effective potentials for the $5g$ electron in element 120 with configuration $8s\,5g$ and element 121 with configuration $8s\,7d\,5g$. Although the change in the potential is relatively small, the $5g$ wave function changes its radius from 25 Bohr units to about 0.6 Bohr units (*101*)

The early onset of the $7p$ shell filling at element 121 can only be understood as a consequence of the large spin-orbit splitting. A discussion of the level structure of the first two elements of the superactinide series, 121 and 122, has been given by *Mann* and *Waber* (*50*) and *Cowan* and *Mann* (*102*). We reproduce in Fig. 20 the level structure of element 121 in comparison to its homolog actinium. The solid lines represent subconfigurations for the individual sets of quantum numbers $|nlj>$ and the barycenter, which is the weighted average taken over the $j$ values, is indicated by a dashed line. Figure 20 nicely shows the large effect of the spin-orbit splitting, which brings a $8p$ electron into the stable atomic ground-state of the first superactinides.

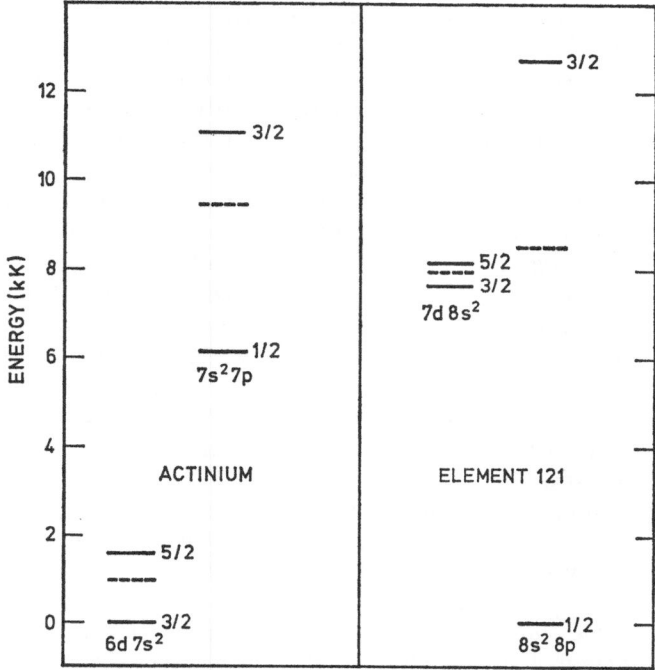

Fig. 20. Calculated term levels of the $ds^2$ and $ps^2$ configurations of actinium and element 121. The dashed lines indicate average energies (50)

The very small binding energies of all electrons in the $8s_{1/2}$, $8p_{1/2}$, $7d_{3/2}$, $6f_{5/2}$ and $5g_{7/2}$ shells makes it plausible that most of these electrons can be oxidized in chemical compounds so that very high oxidation states might be reached in complex compounds. This is in accordance with the trend observed among the lanthanoides, where practically only one oxidation state of 3 is possible, to the actinoides with high oxidation states at the beginning and small oxidation states at the end. Hybridizations of unknown complex character will be possible because the four shells are far enough extended radially and are thus available for hybridization with different angular momenta but nearly the same energy.

With regard to the formation of ionic compounds, it is not too relevant whether the $8p$ or $7d$ shell is occupied in the neutral atom, as studied *in extenso* by *Mann* and *Waber* (50). Instead, the significant question for more ionic compounds is whether in the ions, after all outer $s$, $p$ and $d$ electrons are removed, some $g$ or $f$ electrons will be in frontier orbitals or whether they might be easily excited to an outer electron shell so that they can be removed as well. *Prince* and *Waber* (103) showed that even in the divalent state of element 126 one $g$ electron has changed to an $f$ electronic state. However, the $8s$ electrons are not the first to be removed. Thus, the divalent ions will be expected to act as soft Lewis acids and possibly form covalent complex ions readily. Crystal or ligand fields influence the nature of the hybridization. Details such as directionality of bonds

129

will be determined by the occupation of the frontier orbitals and, of course, by the approaching anion. Only a few calculations for highly ionized states have been done to date. However, from a comparison of the energy eigenvalues of the different shells, it seems that oxidation states may well reach very high values at about, or near element 128 in complex compounds, but that normally these elements will have 4 as their main oxidation state in ionic compounds.

The main reason for this expectation is the observation that the ionization energies increase dramatically with increasing ionization and are soon out of the chemical-energetic range. In addition, there is also a limit imposed by geometrical considerations. In an early paper, *Jørgensen (104)* concludes that element 126 will be mainly tetravalent. The maximal valency will be reduced to 6 at element 132 and in the region of 140 it will be three to four. At the end of the superactinide series, the normal oxidation states are expected to be only 2 because the $6f$ shell is buried deep inside the atom and the $8s$ and $8p_{1/2}$ electrons, which are in frontier orbitals, are bound so strongly that they will be chemically inactive (see Fig. 8); only the $7d$ electrons will be available for bonding. To be more specific, the calculations show that in elements around 156 the shell is nearly full and only two $7d$ electrons which extend radially beyond the $8s$ and $8p_{1/2}$ shell are available. This behavior seems very similar to that of the low oxidation states at the end of the actinoides.

The lanthanoide contraction of about 0.044 Å per element is larger than the actinoide contraction of about 0.03 Å per element, because the $4f$ wave function is less localized than the $5f$ wave function and shrinks more rapidly with increasing nuclear charge. The analogous contraction is expected in the superactinide series. The total effect will be very large because of the 32 electrons, which will be filled in the deep $5g$ and $6f$ shells. From a comparison of the outer-electron wave functions of the lanthanoides, actinoides and superactinides, a contraction of about 0.02 Å per element can be expected, starting with element 121 and continuing to element 154. This can also be seen in Fig. 9.

As a conclusion regarding chemical predictions of these elements, one may say that predictions in this region is somewhat unreliable, first because most of these elements have no homologs; secondly, because we are already well into an unknown region; and third, because relativistic effects have a large influence on these elements. But it is possible to say from the calculations that chemical behavior will be very different for the elements at the beginning, where high oxidation states will readily be reached, and the end where, due to the strong binding of the outermost electrons, the chemical character will be very noble, so that very low oxidation states will be possible.

## 4. The Elements $Z = 155$ to 172 and $Z = 184$

Because there is still a faint possibility that some elements near the magic proton number 164 may have half-lives long enough to permit a chemical study, a discussion of these elements is not purely academic.

The results of the calculations by *Fricke* and *Waber (56)* and *Mann (35)*, which are listed in Table 2, show that formally elements 155 to 164 are the $d$ transition elements of the 8th period.

The relativistic enhancement of the subshells with $j = 1/2$ is so large that in the elements 165 to 168 the $9s$ and $9p_{1/2}$ states will be occupied instead of the $8p_{3/2}$ state. Hence the filling of the $8p_{3/2}$ electrons can occur only in elements 169 to 172. This surprising result makes it possible to give the formal continuation of the periodic table shown in Fig. 21, because there are six $p$ electrons available from two different shells which are energetically very close, so that they will nicely form a "normal" $p$ shell. Therefore, the 9th period will be quite analogous to the 2nd and 3rd periods in the periodic system. This continuation and the differences from the normal expected continuation are discussed below.

**Elements 156 to 164.** In the periods before the 8th period, normally all $d$ and $p$ elements are influenced in their chemical behavior more or less by the outer $s$ electrons. This is no longer true for the $d$ transition elements 155 to 164, where the $8s$ and $8p_{1/2}$ electrons are bound so strongly that they do not participate in the chemical bonding. Fig. 22 shows the outer electronic wave functions of element 164 with the deeply buried $8s$ and $8p_{1/2}$ electrons. This electronic structure is quite similar to that of the $d$ elements of the lower periods, where the outer $s$ electrons are removed. One might therefore argue that, as a first guess, the aqueous and ionic behavior of an $E^{m+2}$ ion of the lower $d$ elements is comparable to an $E^m$ ion of elements 155 to 164 after making allowance for the different ionic sizes and charge. But because the $9s$ and $9p_{1/2}$ states are easily available in 164 for hybridization, the chemical behavior is expected not to be too different from that of the other $d$ elements. *Penneman et al. (71)* gave a very extensive and sophisticated chemical discussion of element 164. They conclude that it would be chemically quite active. In aqueous solution it will be predominantly bivalent, but stronger ligands will form tetra- and hexavalent bonds. Although in its bivalent form it may be compared with lead, it is quite conceivable that tetrahedral $164(CO)_4$ and $164(PF_3)_4$ and linear $164(CN)_2^{-2}$ might be prepared, which would be in rather striking contrast to lead. They find element 164 to be a soft Lewis acid with an Ahrlands softness parameter close to 4 eV, which is very near to the value for mercury. This comparison with Hg agrees well with the position that *Fricke, Greiner* and *Waber (56)* have allocated this element in the periodic table shown in Fig. 21.

*Penneman et al. (72)* arrived at their chemical conclusions from calculations of very high ionization states and a number of semi-empirical formulas, partly discussed in Section III. 2b. They agree with *Fricke et al. (85)* that the metallic form might be quite stable, which again is a result of the deeply buried $8s$ and $8p_{1/2}$ electrons and the readily available $7d$ electrons. This metallic form should have a larger cohesive energy than almost any other element because of the covalent bonding, so that its melting point should be quite high. *Fricke et al. (85)* compare element 164 with element 118 for formal reasons, also. The structures of the valence electrons with a filled outer shell, the ionization energies, the radii, the energy eigenvalues, and the energetic splitting of the filled $d$ states in element 164 and the $p$ shell in element 118 are very similar. Therefore, if element 118, ever becomes available, a detailed experimental study, could also lead to a better understanding of the chemical behavior of element 164.

**Elements 165 and 166.** From the normal continuation of the periodic table one would expect that after the completion of a $d$ shell (at element 164) two elements in the IB and IIB chemical groups should appear.

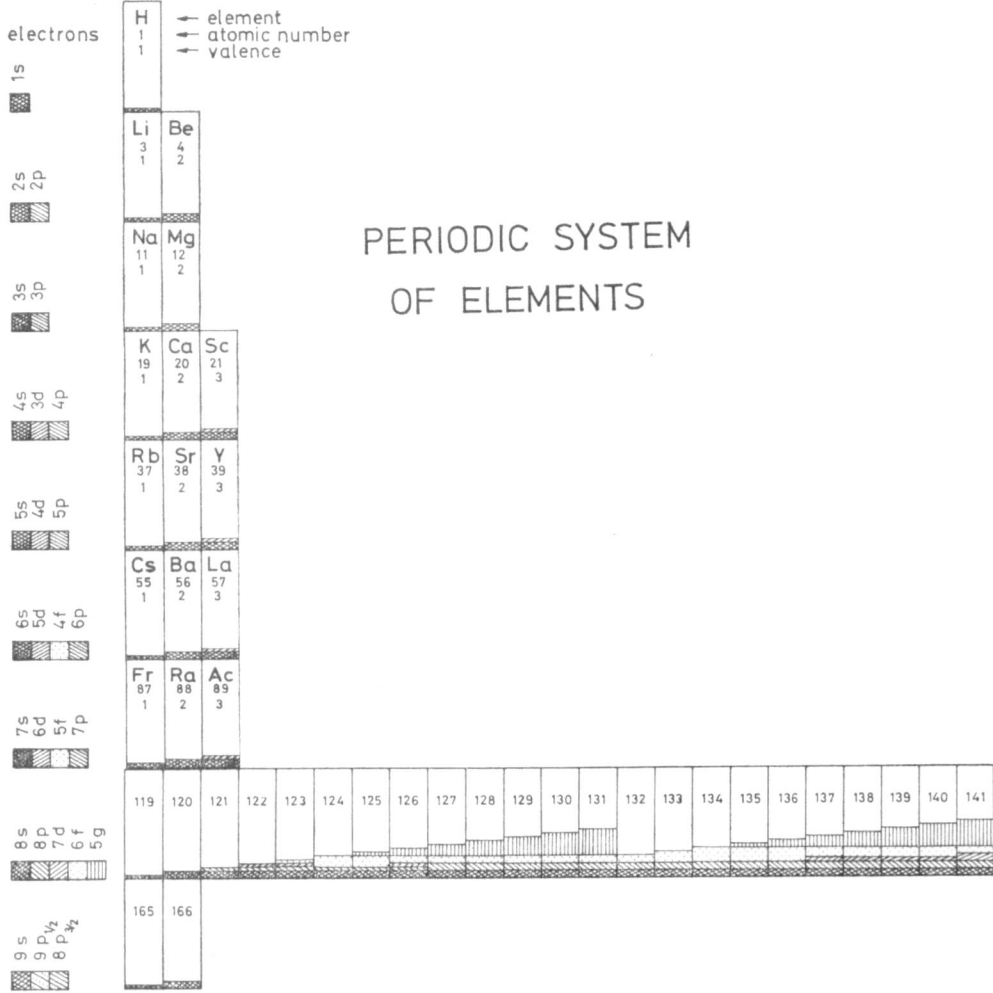

Fig. 21. The periodic system of elements continued up to element 172. The chemical symbols, atomic numbers, and oxidation states are also given. The outer electrons are drawn schematically (56)

In a very formal way this is true, because with the filling of the 9s electrons in elements 165 and 166 there are outer s electrons chemically available. On the other hand, these outer s electrons should be the ones which began with the onset of the period. The 8s electrons are already very strongly bound so that the two 9s electrons which are filled in have to be assumed to define the beginning of a new period. That this interpretation is the correct one can be seen from Fig. 23, where the first ionization energies of the IA and IIA elements are compared with the IB and IIB elements. Because of the result shown in Fig. 23, we certainly include these two elements in the chemical groups of the alkali and alkaline earth

| | | | | | | | | | | | | | | | | | | He 2 0 |
|---|---|---|---|---|---|---|---|---|---|---|---|---|---|---|---|---|---|---|
| | | | | | | | | | | | B 5 3 | C 6 4,2,-4 | N 7 -3,2,5 | O 8 -2 | F 9 -1 | Ne 10 0 | |
| | | | | | | | | | | | Al 13 3 | Si 14 4 | P 15 5,3,-3 | S 16 6,4,-2 | Cl 17 -1,1,5 | Ar 18 0 | |
| Ti 22 4,3 | V 23 5,4,2 | Cr 24 3,6,2 | Mn 25 2,3,4,6,7 | Fe 26 3,2 | Co 27 2,3 | Ni 28 2,3 | Cu 29 2,1 | Zn 30 2 | | Ga 31 3 | Ge 32 4 | As 33 3,-3,5 | Se 34 4,6,-2 | Br 35 -1,1,5 | Kr 36 0 | | |
| Zr 40 4 | Nb 41 5,3 | Mo 42 6,5,3 | Tc 43 7 | Ru 44 3,4,6,8 | Rh 45 3,4 | Pd 46 2,4 | Ag 47 1 | Cd 48 2 | | Jn 49 3 | Sn 50 4,2 | Sb 51 3,5 | Te 52 4,6,-2 | J 53 -1,5,7 | Xe 54 0 | | |

| Ce 58 3,4 | Pr 59 3 | Nd 60 3 | Pm 61 3 | Sm 62 3 | Eu 63 3,2 | Gd 64 3 | Tb 65 3 | Dy 66 3 | Ho 67 3 | Er 68 3 | Tm 69 3 | Yb 70 3,2 | Lu 71 3 | Hf 72 4 | Ta 73 5 | W 74 6 | Re 75 7,4,-1 | Os 76 4,6,8 | Jr 77 3,4,6 | Pt 78 4,2 | Au 79 3,1 | Hg 80 2,1 | Tl 81 1,3 | Pb 82 2,4 | Bi 83 3,5 | Po 84 2,4 | At 85 | Rn 86 0 |
|---|---|---|---|---|---|---|---|---|---|---|---|---|---|---|---|---|---|---|---|---|---|---|---|---|---|---|---|---|
| Th 90 4 | Pa 91 5,4 | U 92 6,5,4,3 | Np 93 5,6,4,3 | Pu 94 4,6,5,3 | Am 95 3,4,5,6 | Cm 96 3 | Bk 97 3,4 | Cf 98 3 | Es 99 3 | Fm 100 3,2 | Md 101 2,3 | No 102 2 | Lw 103 3 | 104 | 105 | 106 | 107 | 108 | 109 | 110 | 111 | 112 | 113 | 114 | 115 | 116 | 117 | 118 |
| 142 | 143 | 144 | 145 | 146 | 147 | 148 | 149 | 150 | 151 | 152 | 153 | 154 | 155 | 156 | 157 | 158 | 159 | 160 | 161 | 162 | 163 | 164 | | | | | | |

| | | | | | | | | | | | | | | | | | | | | | | | 167 | 168 | 169 | 170 | 171 | 172 |
|---|---|---|---|---|---|---|---|---|---|---|---|---|---|---|---|---|---|---|---|---|---|---|---|---|---|---|---|---|

metals (85). This interpretation is also supported by a study of Fig. 10, which shows the calculated radii of these elements resembling those of the elements K and Ca. This classification is, of course, not entirely, satisfactory in every respect because from a more chemical point of view these elements will also show characteristics of the IB and IIB groups because of the underlying 7d shell. Therefore, higher oxidation states than +1 and +2 might readily occur.

**Elements 167 to 172.** Between 167 and 172 the $9p_{1/2}$ and $8p_{3/2}$ electrons will be filled, and it is quite an accident that the energy eigenvalues are so close together (see Fig. 14) that a $p$ shell will occur containing 6 electrons with virtually no

133

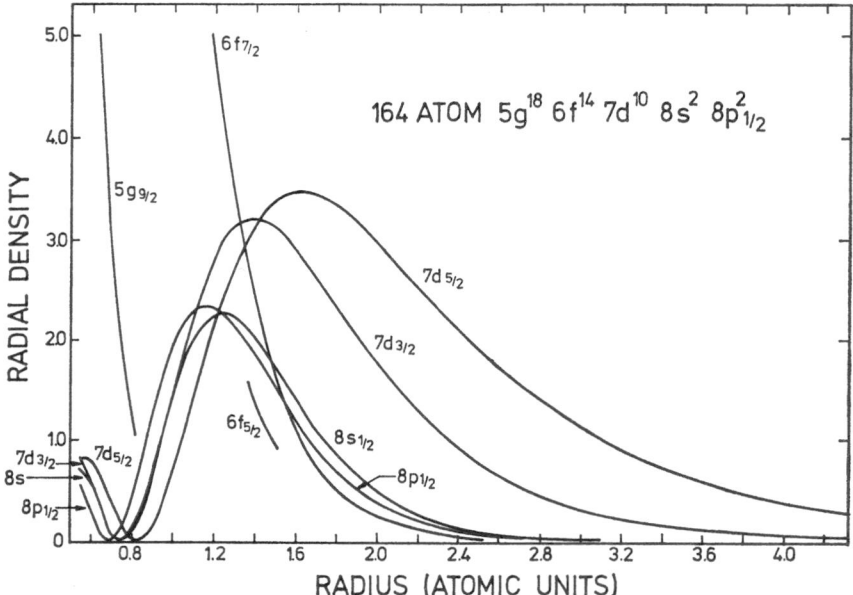

Fig. 22. Radial wave functions of the outer electrons of element 164. The $8s$ and $8p_{1/2}$ electrons are well inside the atom and thus not available for chemical bonding (*55, 71*)

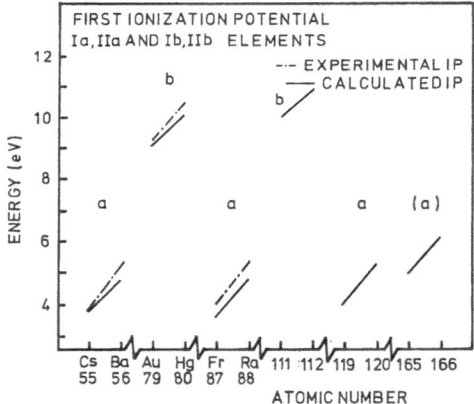

Fig. 23. First ionization energies of the IA, IIA and IB, IIB elements (*85*)

splitting of the subshells but different principal quantum numbers. This situation is analogous to the nonrelativistic $p$ shell in the 3rd period. Therefore, the normal oxidation states of elements 167 to 170 will be 3 to 6. Element 171 is expected to have many possible oxidation states between $-1$ and $+7$, as the halogens do. Here again, the electron affinity will be high enough to form a hydrogen halide like H(171). *Fricke et al.* (*56*) calculated a value for the electron affinity of 3.0 eV, which is as high as the value of J⁻, so that (171)⁻ will be quite a soft base. Element 172 will be a noble gas with a closed $p$ shell outside. The ionization energy of this element, as shown in Fig. 15, is very near to the value of Xe, so that it might be quite similar to this element. The only great difference between Xe and 172 is that element 172 is expected to be a liquid or even a solid at normal temperatures because of its large atomic weight. As indicated in connection with the noble gas 118, element 172 will tend to be a strong Lewis acid and hence compounds with F and O are expected, as has been demonstrated for xenon. In Table 6 the chemical and physical properties of elements 156 to 172 are tabulated.

**Element 184.** *Penneman, Mann* and *Jørgensen (72)* speculated about the chemistry of element 184. This must be regarded as mere speculation because (a) no calculations of nuclear stability had been made up to that time, (b) it seems to be impossible to create this nucleus with any known combination of nuclei, and (c) no calculations of the atomic behavior were available to them. Mann reported that his Dirac-Fock program was unable to go beyond $Z = 176$. Nevertheless, their speculation is quite plausible because after element 172 another extremely long transition series would start with the filling of the 6$g$, 7$f$, and 8$d$ shells. The complication with 6$h$ electronic states might also arise. These loosely bound electrons would mean that it would be very easy to reach very high oxidation states, as stated by *Penneman et al. (71)*. A model calculation of *Fricke* and *Waber (60)* taking into account a phenomenological formulation of quantum-electrodynamical effects makes it possible to extend the Hartree-Fock calculations to even higher elements. They found the ground-state of element 184 to be (164)core $+ 9s^2$ $9p_{1/2}^2 8p_{3/2}^4 6g^5 7f^4 8d^3$. Because of either the small radial extension or the large binding energy, only the 8$d^3$ and 7$f^4$ electrons might be available for chemical bonding. The 10$s$ and 10$p_{1/2}$ electrons do not appear in the ground-state configuration and neither do the 6$h$ electrons. So it seems that the chemical behavior of element 184 is even simpler than that of the early superactinides.

If one goes to higher oxidation states, the occupation of the 6$g$ and 7$f$ shells changes. The main trend is clearly an increase in the number of electrons in the 6$g$ shell. The latter is radially so far inside the atom that these electrons will not be directly available for chemical bonding; instead, the 7$f$ electrons will become involved. From a comparison with uranium, *Fricke* and *Waber (60)* conclude that a $+5$ or $+6$ oxidation state may easily be reached, whereas in aqueous solution the $+4$ oxidation state will be the most stable. Even higher oxidation states seem to be unrealistic because then the electrons from the deeply buried 6$g$ shell, would have had to be removed, and their binding energy increases rapidly with higher ionization. This increase is so great that occupation of the open 6$g$ shell would lead to a deoccupation of the closed 9$s$ and 9$p_{1/2}$ shells, beginning at $184^{+8}$.

This clearly indicates that, here too, in a region of a very long transition series, where many outer electron shells are being filled simultaneously in the neutral atom, the increase in ionization energy is very like what is observed in all other elements. This means that we do not expect extremely high or very unusual oxidation states.

Table 6. Summary of predictions for elements 156 to 172. This table is taken from *Fricke* and *Waber* (85)

| Element | 156 | 157 | 158 | 159 | 160 | 161 | 162 | 163 | 164 | 165 | 166 |
|---|---|---|---|---|---|---|---|---|---|---|---|
| Atomic weight | 445 | 448 | 452 | 456 | 459 | 463 | 466 | 470 | 474 | 477 | 481 |
| Chemically most analogous group | | IIIB | IVB | VB | VIB | VIIB | VIII | VIII | VIII | IA | IIA |
| Outer electrons | $7d^2$ | $7d^3$ | $7d^4$ | $7d^4\,9s^1$ | $7d^5\,9s^1$ | $7d^6\,9s^1$ | $7d^8$ | $7d^9$ | $7d^{10}$ | $9s^1$ | $9s^2$ |
| Most probable oxidation state | +2 | +3 | +4 | +1 | +2 | +3 | +4 | +5 | +2,4,6 | +1,3 | +2 |
| Ionization potential (eV) | 3.7 | 4.8 | 5.7 | 3.8 | 4.1 | 4.5 | 5.4 | 6.0 | 6.8 | 5.0 | 6.1 |
| Metallic radius (Å) | 1.7 | 1.63 | 1.57 | 1.52 | 1.48 | 1.48 | 1.49 | 1.52 | 1.58 | 2.1 | 1.8 |
| Density (g/cm³) | 26 | 28 | 30 | 33 | 36 | 40 | 45 | 47 | 46 | 7 | 11 |

| Element | 167 | 168 | 169 | 170 | 171 | 172 |
|---|---|---|---|---|---|---|
| Atomic weight | 485 | 489 | 493 | 496 | 500 | 504 |
| Chemically most analogous group | IIIA | IVA | VA | VIA | halogen | noble gas |
| Outer electrons | $9s^2\,9p_{1/2}^1$ | $9s^2\,9p_{1/2}^2$ | $9s^2\,9p_{1/2}^2\,8p_{3/2}^1$ | $9s^2\,9p_{1/2}^2\,8p_{3/2}^2$ | $9s^2\,9p_{1/2}^2\,8p_{3/2}^3$ | $9s^2\,9p_{1/2}^2\,8p_{3/2}^4$ |
| Most probable oxidation state | +3 | +4 | +5 | +6 | +7,3,−1 | 0,4,6,8 |
| Ionization potential (eV) | 6.4 | 7.5 | 8.3 | 9.2 | 10.2 | 11.3 |
| Atomic radius (Å) | 1.58 | 1.48 | 1.39 | 1.35 | 1.27 | 1.22 |
| Density (g/cm³) | 17 | 19 | 18 | 17 | 16 | 9 |

# V. Critical Analysis of the Predictions

The predictions of the physical properties and chemical behavior of the super-heavy elements reviewed here can only be a starting point for chemical studies in this region of elements. Nevertheless, it seems very possible that the predictions will not be too far away from reality, at least for the elements up to $Z = 120$. In this region one is still close enough to the part of the periodic system with known elements so that the combination of the simple continuation of trends in the chemical groups together with the results of the very credible calculations, tested in the known part of the elements, will produce quite good predictions about the physical and chemical behavior of superheavy elements. Neither the relativistic effects nor the additional uncertainties will be so large as to create really unexpected new situations.

This statement is rather less true for the elements beyond element 120. In the superactinides we have the unknown chemical behavior of five quite loosely bound and strongly mixed shells together with the unknown chemical behavior of $g$ electrons, and in the region beyond that the structure of the outer electron shells has changed so drastically that only conclusions drawn by analogy can give some idea of the chemical behavior. Even the classification of these elements into chemically analogous groups is not straightforward, so that in constructing the continuation of the periodic table one has to use either more formal or more chemical arguments. In the continuation of the periodic system shown in Fig. 21 we have tried to include both types of argument.

The predictions of the chemical behavior of the elements in the vicinity of the second quasi-stable island are supported only by the calculations within the Hartree-Fock model. The main question in this connection is whether the single-particle Dirac equation is still a good equation for very heavy elements with many electrons and $Z > 137$. That this is true, at least up to $Z = 100$, has been shown by *Fricke, Desclaux* and *Waber* (105). By taking into account the extended nulceus, a formal solution of the Dirac equation is possible up to $Z = 175$. At this point the ls level drops into the continuum of electrons with negative energy. In addition, the interaction between the bound levels and the vacuum becomes so large that the bound electrons and the whole vacuum have to be treated together, as was done theoretically by *Reinhardt et al.* (106). This calculation includes the quantum-electrodynamical effects of vacuum polarization and fluctuation. Also, the effect of retardation in the Coulomb interaction, the magnetic interaction and correlation should be included, in addition, for very large $Z$ elements in an exact manner. All these additional contributions are presently under investigation, but no exact results have been given yet for these very high $Z$ elements. Nevertheless, in the region of the first quasi-stable island these effects are not expected to change the chemical behavior of the elements The only differences one would expect are some small changes in the binding energy of the $j = 1/2$ electrons The changes in the region of the second quasi-stable island of stability might be larger, however. A first heuristic study was done by *Fricke (61)* and *Fricke* and *Waber (60)* by changing the potential near the nucleus, where most of the effects are expected to be maximal, so drastically that the energy eigenvalue of the ls state was raised by about 30% Although this change was very large, the filling of the outer electron shells was only affected *(60)* at the elements 161, 162 and 167 to 172, where the $8p_{3/2}$ and $9p_{1/2}$ shells are filled in the opposite order in the extended calculation. But even this does not change the chemistry significantly. Thus these results may be taken as a first indication that the calculations done within the approximation used in this article can also be quite valuable for the very heavy elements, and that the coupling between the behavior of the inner electrons and the valence electrons is quite small.

# VI. Application of the Chemical Predictions

The predictions of the chemical properties of the superheavy elements discussed in Section IV make it possible to design experiments for their chemical identification should they be produced by heavy-ion bombardement. A few simple preliminary experiments have been performed, utilizing the tandem cyclotron combination at Dubna and the SuperHILAC at Berkeley.

*Flerov, Oganessian* and coworkers *(107)* at Dubna have bombarded uranium with xenon ions and chemically isolated fractions containing the acid-insoluble sulfides of carrier elements, which behave like osmium through bismuth and are the chemical homologs of the superheavy elements 108 to 115. Because the sulfides of these elements are expected to be insoluble in acid solution, it was expected that these superheavy elements would be present in these carrier fractions, but only a few spontaneous fission events were observed in the sulfide fraction, thus giving no evidence of superheavy elements. At Berkeley, *Kratz, Liljenzin* and *Seaborg (94)* have made chemical separations designed to isolate the superheavy elements following bombardment of uranium with argon and krypton ions. Their chemical separations were based on the expected marked tendency of the superheavy elements to form strong complex ions. Especially the elements in the range from 109 to 115 should be soft acceptor ions, as discussed in Section IV, and are expected to form strong complex ions with heavy halide ions such as bromide and iodide, in contrast to the hard acceptors such as the lanthanoide and actinoide ions. They form much weaker complex ions (with essentially electrostatic bonding) with these halide ions. This should provide the means for the separation of such superheavy elements from the lanthanoides and actinoides. As discussed in the section on the elements near 112, these elements are expected to form stable iodide and bromide complex ions, which could actually be used in the separation. Thus the separation problem is reduced to a separation of anions from cations, which can be achieved by (1) cation exchange, (2) solvent extraction with aliphatic amines, and (3) ion exchange.

Keeping in mind that one will have to deal with one atom at a time, one obviously thinks of chromatographic techniques where the separation step is repeated many times and where one can expect even a single atom to behave statistically just like a macroamount of the same substance. The three above-mentioned techniques have been tested in chromatographic columns using radioactive tracers of a variety of elements. The best results were obtained with the cation-exchange procedure. In addition, separation steps for groups of elements have been added by using different volatilities from $HBr/Br_2$ solutions and the different degree of complexing with bromide and chloride ions.

On the basis of these predictions, the chemical separation scheme shown in Fig. 24 has been utilized on thick uranium targets bombarded with argon and krypton ions. With a few exceptions (bromine, iodine, arsenic), no carrier material

CHEMICAL PROCESSING OF HEAVY-ION BOMBARDED
URANIUM TARGETS

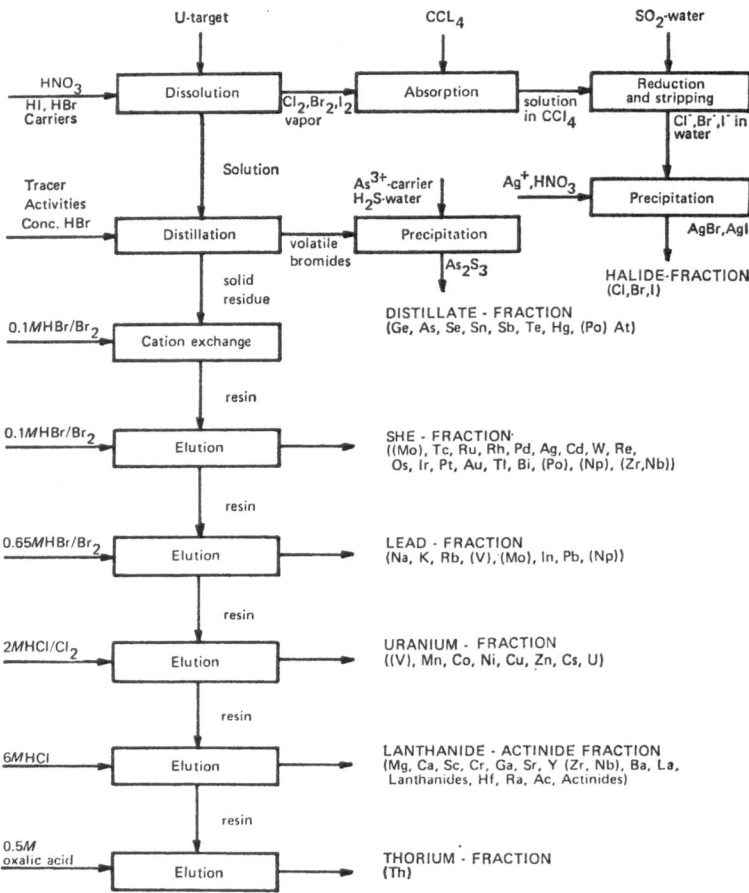

Fig. 24. Chemical processing of heavy-ion bombarded uranium targets as used by *Kratz* and coworkers (*94*). Elements 112 and 114 are expected to appear in fractions separated from the main group of superheavy elements.

was added, *i.e.* carrier-free chemical separations were made leading to the desired thin fractions for measuring $\alpha$ and spontaneous-fission radioactivities. Actually, the details of this separation scheme are such that mercury and lead, presumably acting as prototypes for elements 112 and 114, appear in fractions separated from the main group of superheavy elements and its prototype elements.

When the chemically isolated fractions, which correspond to the superheavy elements, were examined with respect to the detection of possible decay by spontaneous fission and alpha emission, no evidence of such decay was found, thus indicating the absence of superheavy elements. Since this scheme corresponds to a formation cross-section of less than $10^{-35}$ cm$^2$ for an assumed half-life of 100 days, the result is not surprising, because the indicated nuclear reactions are not those considered best suited to the synthesis of superheavy elements.

The expectation that superheavy elements will be detected by chemical and other identification procedures, even with these very small-cross sections, is now shifting to the heavy-ion accelerator laboratory (GSI) in Germany. There is the hope that the use of other heavy ions (including ions up to uranium) and greater beam intensities will lead to the synthesis and identification of superheavy elements. There are also several groups associated with GSI presently developing setups to detect superheavy elements using chemical separation methods similar to those described above as well as phase separations.

Speaking generally, one might say that we are just at the beginning of this field of study. We have a very preliminary idea of what we might expect to find in this large white area of the periodic table. We are sure that a large variety of new phenomena reported from the nuclear physics, atomic physics and chemical points of view will make this field a more and more interesting one.

*Acknowledgements.* I would like to thank Dr. *O. L. Keller* for many years of productive cooperation, advice and help in this field of research. Also I would like to thank Dr. *G. T. Seaborg* for his encouragement and continued advice. I am indepted to Drs. *W. Greiner, J. T. Waber, J. B. Mann, R. A. Penneman* and *C. K. Jørgensen* for the many discussions and communications that made this work possible.

# VII. References

1. *Oganesian, Y. T., Tretyakov, Y. P., Iljinov, A. S., Demin, A. G., Pleve, A. A., Tretja-kova, S. P., Plotko, V. M., Ivanov, M. P., Dalinov, N. A., Korotkin, Y. S., Flerov, G. N.:* Dubna report JINR-D7-8099.
2. *Ghiorso, A., Hulet, E. K., Nitschke, K. M., Alonso, J. R., Lougheed, R. W., Alonso, C. T., Nurmia, M., Seaborg, G. T.:* Phys. Rev. Letters, *33*, (1974).
3. see for example *Seaborg, G. T., Katz, J. J.:* The actinide elements, national nuclear energy series. New York: McGraw-Hill 1949.
4. *Seaborg, G. T.:* Actinides Rev. *1*, 1 (1968).
5. *Seaborg, G. T.:* Ann. Rev. Nucl. Sci. *18*, 53 (1968).
6. *Seaborg, G. T.:* Man-made transuranium elements Englewood Cliffs, N. J.: Prentice-Hall 1964; see also Ref. (*5*).
7. *Seaborg, G. T.:* The transuranium elements. New Haven. Yale University Press 1958. — *Katz, J. J., Seaborg, G. T.:* The chemistry of the actinide elements. New York: John Wiley and Sons 1957.
8. *Cunningham, B. B.:* Ann. Rev. Nucl. Sci. *14*, 323 (1964).
9. *Asprey, L. B., Penneman, R. A.:* Chem. Eng. News *45*, 75 (1967).
10. *Keller, C.:* The chemistry of transuranium elements. Weinheim: Verlag Chemie 1971.
11. *Silva, R. J.:* MTP Int. Rev. Sci., Series I, Inorganic Chemistry, Vol. *8*, 71. London: Butterworth 1972.
12. *Wheeler, J. A.:* Niels Bohr and the development of physics, p. 163. London: Pergamon 1955. — *Wheeler, J. A.:* Proc. Int. Conf. Peaceful Uses of Atomic Energy, Geneva *2*, 155 and 220. New York: United Nations 1956.
13. *Scharff-Goldhaber, G.:* Nucleonics *15*, 122 (1957).
14. *Werner, F. G., Wheeler, J. A.:* Phys. Rev. *109*, 126 (1958).
15. *Petrzhak, K. A., Flerov, G. N.:* Soviet Phys. Usp. (English Transl.) *4*, 305 (1961).
16. *Meldner, H., Röper, G.:* unpublished result quoted in Ref. (*18*) (1965).
17. *Meldner, H.:* Arkiv Fysik *36*, 593 (1967).
18. *Myers, W. D., Swiatecki, W. J.:* Nucl. Phys. *81*, *1* (1966).
19. Detailed reviews of this field have been given in: *Nix, J. R.:* Ann. Rev. Nucl. Sci. *22*, 65 (1972). — *Nix, J. R.:* CERN-Report 70—30, page 605 (1970). — *Johannson, T., Niels-son, S. G., Szymanski, Z.:* Ann. Phys. (Paris) *5*, 377 (1970).
20. *Nielsson, S. G., Nix, J. R., Sobiczewski, A., Szymanski, Z., Wyceck, S., Gustafson, C., Möller, P.:* Nucl. Phys. *A 115*, 545 (1968).
21. *Tsang, C. F., Nielsson, S. G.:* Nucl. Phys. *A 140*, 289 (1970).
22. *Muzychka, A., Pashkwich, V. V., Strutinski, V. M.:* Soviet J. Nucl. Phys. (English Transl.) *8*, 417 (1968).
23. *Mosel, U., Greiner, W.:* Z. Physik *222*, 261 (1969).
24. *Meldner, H.:* Phys. Rev. *178*, 1815 (1969).
25. *Köhler, H. S.:* Nucl. Phys. *A 162*, 385 (1971) and *A 170*, 88 (1971).
26. *Vautherin, D., Veneroni, M., Brink, D. M.:* Phys. Letters *33 B*, 381 (1970).
27. *Rouben, B., Pearson, J. M., Saunier, G.:* Phys. Letters *42 B*, 385 (1972).
28. *Sobiczewski, A., Krogulski, T., Blocki, J., Scymanski, Z.:* Nucl. Phys. *A 168*, 519 (1971).
29. *Seaborg, G. T.:* in Ref. (*30*) page 5.
30. Proc. *Robert A. Welch* Found. Conf. XIII — The Transuranium Elements, ed. *W. O. Milligan.*
31. For a summary, see: *Pauli, H. C.:* Phys. Rep. *7*, 35 (1973).
32. *Herrmann, G.:* MTP Intern. Rev. Sci. Ser. 2 Radiochemistry, p. 221. London: Butter-worth 1975. — *Herrmann, G.:* Nobel Symposium on Superheavy Elements. Ronneby: Sweden, June 1974.

33. *Fiset, E. O., Nix, J. R.:* Nucl. Phys. *A 193,* 647 (1972).
34. *Grumann, J., Mosel, U., Fink, B., Greiner, W.:* Z. Physik *228,* 371 (1969). — *Grumann, J., Morović, T., Greiner, W.:* Z. Naturforsch. *26 a,* 643 (1971).
35. *Mann, J. B.:* to be published by the USSR Academic of Sciences, Institut of the History of Natural Sciences and Technology in their 1975 Centenary Volume of the Discovery of Gallium predicted by *D. I. Mendeleev.*
36. *Madelung, E.:* Die Mathematischen Hilfsmittel des Physikers, Appendix 15. Berlin: Springer 1936. This result was not published before 1936.
37. *Gol'danskii, V. I.:* Priroda *2,* 19 (1969).
38. *Chaikkorskii, A. A.:* Soviet Radiochem. (English Transl.) *12,* 771 (1970).
39. *Taube, M.:* Nukleonika *12,* 309 (1967).
40. An excellent description of the Hartree and Hartree-Fock method is given by *Slater, J. C.:* Quantum theory of atomic structure, Vol. I, II. New York: McGraw-Hill 1960.
41. *Hartree, D. R.:* The calculation of atomic structure. New York: John Wiley and Sons 1957.
42. *Herman, F., Skillman, S.:* Atomic structure calculations. Englewood Cliffs, N. J.: Prentice-Hall 1963.
43. *Grant, I. P.:* Advan. Phys. *19,* 747 (1970).
44. *Larson, A. C., Waber, J. T.:* J. Chem. Phys. *48,* 5021 (1968) and Los Alamos Scientific Lab. Report LA-DC-8508.
45. *Froese-Fischer, C.:* Comp. Phys. Commun. *1,* 151 (1969); *4,* 107 (1972).
46. *Slater, J. C.:* Phys. Rev. *81,* 385 (1951).
47. *Dirac, P. A. M.:* Proc. Roy. Soc. (London) *A 117,* 610 (1928); *A 118,* 351 (1928).
48. Such programs were developed by: *Mann, J. B.:* unpublished. — *Desclaux, J. P.:* Comp. Phys. Commun., *9,* 31 (1975).
49. *Coulthard, M. A.:* Proc. Phys. Soc. *91,* 44 (1967).
50. *Mann, J. B.:* J. Chem. Phys. *51,* 841 (1969). — *Mann, J. B., Waber, J. T.:* J. Chem. Phys. *53,* 2397 (1970).
51. *Mann, J. B., Waber, J. T.:* Atomic Data *5,* 201 (1973).
52. *Liberman, D. A., Cromer, D. T., Waber, J. T.:* Comp. Phys. Commun. *2,* 107 (1971).
53. *Desclaux, J. P.:* Comp. Phys. Commun. *1,* 216 (1969). — *Tucker, T. C., Roberts, L. D., Nestor, C. W., Carlson, T. A., Malik, F. B.:* Phys. Rev. *174,* 118 (1968).
54. *Waber, J. T., Cromer, D. T., Liberman, D.:* J. Chem. Phys. *51,* 664 (1969).
55. *Fricke, B., Greiner, W.:* Phys. Letters *30 B,* 317 (1969).
56. *Fricke, B., Greiner, W., Waber, J. T.:* Theoret. Chim. Acta (Berlin) *21,* 235 (1971).
57. *Mann, J. B.:* page 431 in Ref. *(30).*
58. *Malý, J., Hossonnois, U.:* Theor. Chim. Acta (Berlin) *28,* 363 (1973).
59. *Fricke, B., Waber, J. T.:* J. Chem. Phys. *56,* 3726 (1972).
60. *Fricke, B., Waber, J. T.:* J. Chem. Phys. *57,* 371 (1972).
61. *Fricke, B.:* Lett. al Nuovo Cimento *2,* 293 (1969).
62. *Averill, F. W., Waber, J. T.:* Submitted to J. Chem. Phys.
63. *Slater, J. C.:* J. Chem. Phys. *41,* 3199 (1964).
64. *Koopmanns, T.:* Physica *1,* 104 (1933).
65. *Waber, J. T.:* page 353 in Ref. *(30).*
66. *Nugent, L. J., Van der Sluis, K. L., Fricke, B., Mann, J. T.:* Phys. Rev. *A 9,* 2270 (1974). — Papers by *J. B. Mann* and *B. Fricke* in the: International Symposium on the electronic Structure of the Actinides, Argonne, Oct. 1974.
67. *Fricke, B., Waber, J. T.:* J. Chem. Phys. *56,* 3246 (1972).
68. *Jørgensen, C. K.:* Angew. Chem. *85,* 1 (1973).
69. *Pauling, L.:* The nature of the chemical bond, 3rd ed. Cornell Univ. Press 1960.
70. *Phillips, C. S., Williams, R. J.:* Inorganic chemistry. Oxford: University Press 1966.
71. *Day, M. C., Selbin, J.:* Theoretical inorganic chemistry. New York: Van Nostrand Reinhold 1969.
72. *Penneman, R. A., Mann, J. B., Jørgensen, C. K.:* Chem. Phys. Letters *8,* 321 (1971).
73. *Jørgensen, C. K.:* Modern aspects of ligand field theory, pp. 263, 432. Amsterdam: North-Holland 1971. — *Jørgensen, C. K.:* Chimia *23,* 292 (1969).

B. Fricke

74. *David, F.:* Institut de Physique Nucleaire, Division de Radiochemie, Report RC-71-06 (1971).
75. *Bächmann, K., Hoffmann, P.:* Radiochim. Acta *15*, 153 (1973).
76a. *Hoffmann, P.:* Radiochim. Acta *19*, 69 (1973).
76b. *Hoffmann, P.:* Radiochim. Acta *20*, in press (1974).
76c. *Hoffmann, P.:* Radiochim. Acta *17*, 169 (1972).
77. *Eichler, B.:* Dubna Report JINR P12-7767 (1974).
78. *Keller, O. L., Burnett, J. L., Carlson, T. A., Nestor, C. W.:* J. Phys. Chem. *74*, 1127 (1970).
79. *Halliwell, H. F., Nyburg, S. C.:* Trans. Faraday Soc. *59*, 1126 (1963).
80. *Delahay, P.:* New instrumental methods in electrochemistry, p. 268. New York: Interscience 1954.
81. *Johnson, D. A.:* Some thermodynamic aspects of inorganic chemistry, p. 86 ff. Cambridge: University Press 1968.
82. *Ahrland, S.:* Chap. 5 in Structure and Bonding (*C. K. Jørgensen*, ed.). Berlin-Heidelberg-New York: Springer 1968.
83. *Pearson, R. G.:* J. Chem. Educ. *45*, 581 and 643 (1968).
84. *Klopman, G.:* J. Am. Chem. Soc. *90*, 223 (1968).
85. *Fricke, B., Waber, J. T.:* Actinides Rev. *1*, 433 (1971).
86. *Zvara, I., Chuburkov, Y. T., Tsaletka, R., Tsarova, T. S., Shelevski, M. R., Shilov, B. V.:* Atomnaya Energiya *21*, 83 (1969). — *Zvara, I. et al.:* J. Inorg. Nucl. Chem. *32*, 1885 (1970).
87. *Zvara, I. et al.:* Soviet Radiochem. (English Transl.) *9*, 226 (1967). — *Zvara, I.:* 24th Int. Conf. on Pure and Applied Chemistry Hamburg, Sept. 1973.
88. See, for example, the discussion of the *d* elements by *Cunningham, B. B.*, p. 307, ref. 30.
89. *Penneman, R. A., Mann, J. B.:* J. Inorg. Nucl. Chem., in press.
90. *Häissinsky, M.:* Radiochem. Radioanal. Letters *8*, 107 (1971); J. Chim. Phys. *5*, 845 (1972).
91. *Seaborg, G. T.:* 24th Int. Congr. on Pure and Applied Chemistry Hamburg, Sept. 1973.
92. *Keller, O. L.:* to be published in the same book as Ref. (*35*).
93. *Keller, O. L., Nestor, C. W., Carlson, T. A., Fricke, B.:* J. Phys. Chem. *77*, 1806 (1973).
94. *Kratz, J. V., Liljenzin, J. O., Seaborg, G. T.:* Inorg. Nucl. Chem. Letters *10*, 951 (1974).
95. *Grosse, A. V.:* J. Inorg. Nucl. Chem. *27*, 509 (1965).
96. *Cunningham, B. B.:* Ref. 140 in Ref. (*5*).
97. *Jørgensen, C. K., Häissinsky, M.:* Radiochem. Radioanal. Letters *1*, 181 (1969).
98. *Hoffmann, P.:* Radiochem. Radioanal. Letters *14*, 207 (1973).
99. *Keller, O. L., Nestor, C. W., Fricke, B.:* J. Phys. Chem. *78*, 1945 (1974).
100. *Smith, G. P., Davis, H. L.:* Inorg. Nucl. Chem. Letters *9*, 991 (1973).
101. *Griffin, D. C., Andrew, K. L., Cowan, R. D.:* Phys. Rev. *117*, 62 (1969).
102. *Cowan, R. D., Mann, J. B.:* The Atomic Structure of Superheavy Elements, in Atomic Physics 2 (*P. G. H. Sandras*, ed.). New York: Plenum Press 1971.
103. *Prince, M., Waber, J. T.:* published in Ref. (*30*).
104. *Jørgensen, C. K.:* Chem. Phys. Letters *2*, 549 (1968).
105. *Fricke, B., Desclaux, J. P., Waber, J. T.:* Phys. Rev. Letters *28*, 714 (1972).
106. *Reinhardt, P. G., Greiner, W., Ahrenhövel, H.:* Nucl. Phys. A *166*, 173 (1971).
107. *Flerov, G. N.:* Paper given at the Int. Conf. on Reactions between Complex Nuclei, Nashville, June 1974. — *Zvara, I.:* 24th Int. Cong. on Pure and Applied Chem., Hamburg, Sept. 1973.
108. *Chaikhorski, A. A.:* Radiochem. (USSR) *14*, 122 (1972).
109. As a guidance to this field see: *Mohler, P. H., Stein, H. J., Armbruster, P.:* Phys. Rev. Lett. *29*, 827 (1972); — *Mohler, P. H.:* Proceedings of the IV. Int. Conf. Atomic Physics, Heidelberg, Germany, July 1974, in press.

Received December 4, 1974

# Structure and Bonding: Index Volume 1-21

**B. Volmert**

# POLYMER CHEMISTRY

Translated from the German by E. H. Immergut,
New York
With 630 figures. XVII, 652 pages. 1973
Cloth DM 72,–; US $29.60
ISBN 3-540-05631-9
Prices are subject to change without notice

This book gives a comprehensive coverage of the synthesis of polymers and their reactions, structure, and properties. The treatment of the reactions used in the preparation of macromolecules and in their transformation into cross-linked materials is particularly detailed and complete. The book also gives an up-to-date presentation of other important topics, such as enzymatic and protein synthesis, solution properties of macromolecules, polymer crystallization, and properties of polymers in the solid state.

The content and presentation of Professor Vollmert's book is more encompassing than most existing treatises, and its numerous figures and tables convey a wealth of data, never, however, at the expense of intellectual clarity or educational value.

The presentation is mainly on a fundamental and general level and yet the reader—student or professional—is gradually and almost casually introduced to all important natural and synthetic polymers. Complicated phenomena are explained with the aid of the simplest available examples and models in order to ensure complete understanding. However, the reader is also encouraged to think for himself and even to criticize the author's point of view.

All of the chapters have been revised and enlarged from the German edition, and many of the sections are entirely new.

**Contents**
Introduction. — Structural Principles. — Synthesis and Reactions of Macromolecular Compounds. — The Properties of the Individual Macromolecule. — States of Macromolecular Aggregation.

## Springer-Verlag
## Berlin Heidelberg New York

NMR — Basic Principles and Progress
Grundlagen und Fortschritte
Editors: P. Diehl, E. Fluck, R. Kosfeld

**Volume 6**

P. Diehl, H. Kellerhals, E. Lustig:
Computer Assistance in the Analysis of High-Resolution NMR Spectra
11 figures. III, 96 pages. 1972. Cloth DM 48,—; US $19.70
ISBN 3-540-05532-0

**Volume 7**

56 figures. V, 153 pages. 1972. Cloth DM 78,—; US $32.00
ISBN 3-540-05687-4

**Volume 8**

C. Richard, P. Granger:
Chemically Induced Dynamic Nuclear and Electron Polarizations —
CIDNP and CIDEP
26 figures. II, 127 pages. 1974. Cloth DM 58,—; US $23.80
ISBN 3-540-06618-7

Prices are subject to change without notice

# Springer-Verlag
# Berlin Heidelberg New York